GREENWAY :

绿道

——哈尔滨向阳镇发展战略规划
Development Strategic Planning of Xiangyang Town in Harbin

〔西〕胡安·布斯盖兹（Joan Busquets）

高岩 周艳莉 周雪瑶

著

中国建筑工业出版社

编委会

主　　任：胡安·布斯盖兹（Joan Busquets）

副 主 任：高　岩　周艳莉　周雪瑶　弗朗西斯·巴克（Francesc Baqué）

顾　　问：栾志成　李四川　谭乐伟　赵　罡　司炳春　孙凤玲　徐　强
　　　　　王宇虹　曹鸿雁　陈贵州　孙汝峰　刘佰军　杨岩松　马成宇
　　　　　刘　刚

编　　委：吴昊华　李春梅　于戌申　王嘉瑞　刘向宇　韩　杰　于艳辉
　　　　　郑永梅　宋　磊　万　宁

编写单位：西班牙BLAU景观·建筑·城市规划设计事务所
　　　　　哈尔滨市城乡规划设计研究院

这本书主要研究了哈尔滨城市及其郊区的都市特性和发展演变，这其中也包括了向阳镇。

The book is about the research on urban qualities and transformation of Harbin and its metropolitan development, including Xiangyang.

对于我们的研究来说，要充分理解哈尔滨因其地理位置所具有的城市特性：当地气候条件和城市历史让这座城市独具风格。

It is important to understand Harbin's urban specificity because of its location in China: the climate condition and its historical development makes the city quite unique.

我们的研究方向主要针对哈尔滨都市圈的规划设计。那么为了让城市规划落到实处，我们必须提出正确的问题，理解现状条件，并且基于现状提出能够刺激城市发展的新战略。因此，我们需要将多个不同层面的研究同时结合到一起：研究基础设施的变化（公路、地铁、水利等），还有为城市未来的发展规划中远期的经济结构。

Research exercise for Harbin is oriented towards designing the metropolitan city. Then, urban planning must be specific by addressing the right questions, understanding current condition, and preparing new strategies that can introduce development for this metropolitan reality. To achieve it, it's necessary to combine several scales at the same time;meaning: studying infrastructural changes (like motorways, metro, water provision, ...), but also, defining economic structures that can make this reality stronger and richer in the mid-term run.

在这样的前提下，我们采用了一种全新的规划方式，以突破现今很多国际大都市城市规划仍在采用的"功能主义模型"。这种新思维鼓励我们去探索东哈尔滨如何能够成为在中国东北黑龙江流域重新定义城市与农业关系的最佳目的地。

Within this frame it seems quite interesting to develop a new approach trying to overcome the "functionalistic model" still operating in many large Metropolis. This new attitude is inviting us to explore how the East of Harbin can be the right location to redefine the relationship between City and Agriculture in the Heilongjiang River region within Northeastern China.

我们可以制定一个与众不同的"城郊规划策略",以发挥东哈尔滨（向阳镇及其广义周边地区）的农业潜力,在阿什河流域下游定义一个名为"郊野绿道"的农业公园。这一建立大型的农业公园的规划决策,将为哈尔滨郊区带来全新的都市结构,即对于整个地区来说全新的城乡形态;同时,这一策略可以说是过去几十年世界其他大城市的规划策略在本地农业区的新型应用。

It is possible to define a different "Metropolitan strategy", acknowledging the agricultural potential of the East—considering Xiangyang at large, defining an Agricultural Park as will be described later as UCAP after the Ashi corridor. For such a reason, the strategy of a large Agricultural Park could be an advisable planning decision that implies a new urbanistic structure at the level of Metropolis, meaning a new formal morphology for the overall territory; at the same time this strategy can represent a spinoff for the Agriculture in the region following the patterns of other large cities across the globe in the last decades.

胡安·布斯盖兹教授
Prof. Joan Busquets

目 录
CONTENTS

引言
INTRODUCTION

本规划研究并提出了哈尔滨空间转变发展的新框架。建立起这一系列的空间系统和设计导则对于整个地区的发展是至关重要的。这不是一个宏观的规划，而是农业、经济、空间、活动、建筑、开放空间和管理方式的新组合，它的实现将整个地区的内在品质凸显并加以利用。

本规划既是连贯性的框架，也是多尺度开发可以组合在一起的蓝图。这种规模的规划需要始终具有一定的灵活性和弹性，以允许新的发展理念可以融入，并在适当的时间作出新的决策。

向阳镇是21世纪哈尔滨新型城镇化发展的先行者，我们将它作为哈尔滨空间新框架的试点地。这里是21世纪城市化需要发展的地方，也许我们可以用向阳镇作为它的试点计划。

The project is the "Frame" for a large transformation which is envisioned for the area. It seems crucial to be able to establish a quite coherent system of spaces and initiatives which may channelize the success of it. It's not a piece of macro-architecture but a combination of agricultural and economic space, different events, architecture, open spaces and management initiatives that may carried it out.

Intrinsic qualities of the place must be enhanced to take the benefit of it.

Project is the frame for coherence but also is the canvas where multiple scale development can fit together. Project at this scale needs always a certain flexibility to allow new input to proceed and new decisions to be taken in the right time. Here is where the Urbanism of the 21st century needs to be developed;perhaps we may use Xiangyang Project as pilot initiative for it.

场所的内在品质必须被加强和利用
INTRINSIC QUALITIES OF THE PLACE MUST BE ENHANCED TO TAKE THE BENEFIT OF IT

规划雄心

SUMMARY OF MAIN AMBITIONS FOR THE PROJECT

历史上哈尔滨东部区域包括向阳镇一直处于边缘化状态，经济实力薄弱。向阳镇以北和以东的区域有相似的处境，然而向阳镇以南的部分靠近阿什河的小镇相对有着更雄厚的经济实力。例如，在料甸镇有哈尔滨的第一个特色满族村，同时非常靠近红星水库景区；成高子镇现有运营非常成功的伏尔加庄园项目；另外在阿城保留有金代遗址，同样比向阳镇更具吸引力和经济实力。

由于长江路、哈牡高铁的建设与开通，改变了向阳镇的经济区位，向阳镇与哈尔滨市的联系更加紧密，在哈尔滨东部各镇中，向阳镇占据了相当重要的地理位置。

虽然向阳镇与哈尔滨东部相邻的料甸镇、成高子镇伏尔加庄园相比经济实力较弱，与哈牡高铁沿线上的旅游强镇亚布力旅游滑雪度假胜地、一面坡特色小镇、帽儿山镇、玉泉镇等相比旅游吸引力较弱，但本次规划最终目的是通过搭建哈尔滨空间发展新框架，同时利用旅游快线哈牡高铁起点与终点分别临近哈尔滨市热门景点及镜泊湖的牡丹江的溢出效应及它们所具有的强大吸引力，提升东部地区及向阳镇的内在品质并加以利用，从而增强东部地区及向阳镇的旅游吸引力，使之成为东部地区及向阳镇经济发展的新引擎。

Currently, Xiangyang plays a significant role in East Harbin region, mainly due to its location, where Changjiang Road also Hamu highspeed rail and metro line 2, pass through the town. But most of the towns in east region developed slowly in the past and plays an insignificant economic role in Harbin. The towns that locate on the east and north of Xiangyang have similar economic condition as Xiangyang. But many towns which locate on the south of Xiangyang have better quality. For example, Harbin's first Manchu characteristic village locates in Liaodian town, which also is very close to Hongxing Reservoir Scenery. The Volga Manor in Chenggaozi Town is also a very successful project. And Acheng, which has a famous historical site of Jin Dynasty, also has better economic condition than Xiangyang.

Xiangyang plays a minor role among all the towns connected by the Hamu highspeed rail. The start point, Harbin is no doubt an important city. The terminal point, Mudanjiang city is also a very significant destination, which is close to Jingpo Lake. And Currently the tourist towns – Yabuli Ski Resort, Yimianpo Characteristic Town, Maoer Mountain, Harbin Forest Zoo, Historical Site of Jin Dynasty and Liaodian Manchu town, all of above are more attractive and popular destinations than Xiangyang. Nevertheless, UCAP Project –as will be presented– can help in creating a new potential for the whole region.

11

现状概况
CURRENT SITUATION

地理位置
GEOGRAPHIC LOCATION

　　向阳特色旅游镇位于哈尔滨市香坊区东北部，阿什河东岸，距中心城区约 14km，约 15 分钟车程；距四环路仅 6km。向阳镇镇域北与哈尔滨市道外区团结镇、民主镇接壤，西、南与香坊区成高子镇、阿城区料甸镇为邻，东与道外区永源镇相连。从旅游区位上看其位于哈尔滨东部旅游带上，游玩 4A 级亚布力旅游度假区途经此地，毗邻享有盛名的伏尔加庄园。尤其是哈牡高铁的通车，缩短了向阳特色旅游镇与主城区、亚布力旅游度假区、牡丹江镜泊湖等旅游胜地的空间距离。

Xiangyang characteristic tourism town is located in the northeast of Xiangfang District, Harbin City, on the East Bank of Ashi River, about 14 kilometers from the central city, about 15 minutes by car;only 6 kilometers from the Fourth Ring Road. Xiangyang Town is bordered by Tuanjie Town and Democracy Town in Daowai District of Harbin in the north, Chenggaozi Town in Xiangfang District in the West and Liaodian Town in Acheng District in the south, and Yongyuan Town in the east. From the point of view of tourism location, it is located in the eastern tourism belt of Harbin. It passes through the Yabuli tourist resort of Grade 4A, adjacent to the famous Volga Manor. Especially, the opening of Hamu highspeed railway shortens the space distance between Xiangyang characteristic tourist town and the main urban area, Yabuli tourist resort, Mudanjiang Jingpo Lake and other tourist resorts.

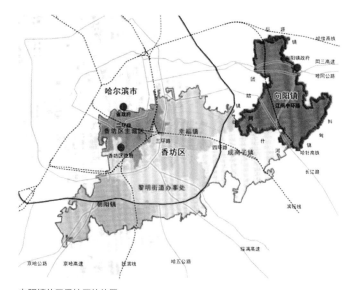

向阳镇位于哈尔滨市的位置
Xiangyang Town is located in Harbin

向阳镇位于香坊区的位置
Xiangyang Town is located in Xiangfang District

区位分析
URBAN CONTEXT

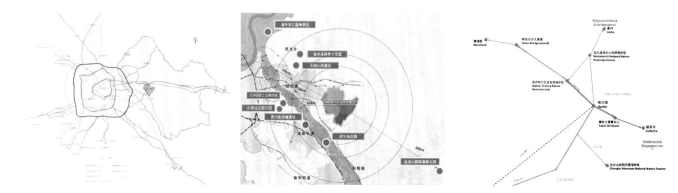

规划区域距离四环 6km，是城市的后花园。从旅游区位上看位于哈尔滨东部旅游带上。
The planned area is 6 km from the Fourth Ring and is the back garden of the city. From the tourist location, it is located on the tourist belt in the east of Harbin.

自然概况
NATURAL ENVIRONMENT

地形地貌：向阳镇地处半丘陵地带，地形高差 69m，平坦地区标高为 121~135m，丘陵地区标高 150~190m；"七山一水二分田"概括了镇域的地貌形态，它是哈尔滨市主城区周边乡镇唯一处在黑土漫岗上的乡镇。

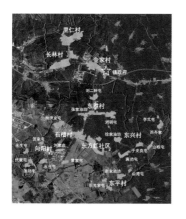

自然资源：镇域内有阿什河、小黄河、杨洪业大壕等沟河，阿什河流经镇内西部，全长 3km 左右；西泉眼水库的灌渠纵贯南部，河流沿岸水草资源丰富；2009 年在东兴村西齐家屯发现了富含矿物质的白垩系地下水，其水质达到国家饮用天然矿泉水标准，该水为含偏硅酸锶型矿泉水，储量丰富，单眼井涌水量每天达 2467t，若开发为饮用天然矿泉水，可供上百万人饮用。

历史沿革
HISTORICAL EVOLUTION

向阳镇特色旅游镇涉及 5 个行政村和 1 个社区。其中，历史记载的主要是东兴村、东平村。

东北是满族的龙兴之地。清顺治元年（公元 1644 年），满族八旗兵丁和汉军八旗部队"从龙入关"，几乎全部从东北关外迁拨入关中平原地区，东北地区人员稀少，仅有一些触犯刑律者"流人"被发配至"极边寒苦"的东北。至清末，东北才正式成为中国主要的农业生产基地。

乾隆 21—23 年（1755—1757 年）间，京旗移民至阿城，由于移民人员过多，阿城安置出现困难，部分移民安置在现在的东兴、东平村位置，形成满族旗屯，至今已有 260 多年历史，即东兴、东平村。旗屯形成初期，屯中主要有满族八大姓氏"佟、关、索、马、齐、富、那、郎"，现在村中大姓为"张"姓及"关"姓。"前关家""后关家"等自然屯地名均由此而来。

Topography and landform: Xiangyang Town is located in the semi-hilly area, with a height difference of 69 meters, elevation of 121-135 meters in the flat area and 150-190 meters in the hilly area. "Seven mountains, one water and two fields" summarizes the geomorphological form of the town, which is the only town in the vicinity of the main urban area of Harbin on the black soil hillock.

Natural resources: There are Ashi River, Xiaohuang River, Yanghongye trench and other ditches in the town area. Ashi River runs through the western part of the town, with a total length of about 3 kilometers. The flax irrigation canal of Xiquanyan Reservoir runs through the south, and there are abundant water and grass resources along the river. In 2009, mineral-rich Cretaceous groundwater was found in Xiqijiatun, Dongxing Village. Its water quality meets the national drinking natural mineral water standard. The water is strontium metasilicate-type mineral water with abundant reserves. The water inflow from a single well reaches 2467 tons per day. If developed as drinking natural mineral water, it could be used for millions of people to drink.

Xiangyang characteristic tourism town involves five administrative villages and one community. Among them, the main historical records are Dongxing Village and Dongping Village.

Northeast China is the place of Manchu's Longxing. In the first year of Shunzhi in the Qing Dynasty (AD 1644), Manchu Eight-Banner soldiers and Han Eight-Banner Army troops "entered the Guanzhong Plain from the dragon". Almost all of them moved to the Guanzhong Plain from outside the Northeast Pass. There were few people in the Northeast, and only some of them who violated the criminal law "exiled" frequently distributed in the Northeast with "extreme cold and bitter". It was not until the late Qing Dynasty that Northeast China became the main agricultural production base in China.

During the period of 21-23 years (1755-1757) of Qianlong reign, Beijing Banner immigrated to Acheng. Because of the excessive number of immigrants, the settlement of Acheng was difficult. Some immigrants were settled in the present Dongxing and Dongping villages, forming the Manchu Qitun, which has a history of more than 260 years. Improvised Dongxing and Dongping Village. In the early days of Qitun formation, there were eight surnames of Manchu in Tunzhong, namely "Tong, Guan, Suo, Ma, Qi, Fu, Na and Lang". Now the surnames of the village are "Zhang" and "Guan". Natural place names such as "Qianguanjia" and "Houguanjia" all come from this.

人口状况
POPULATION STATUS

人口规模：向阳镇特色旅游镇规划范围内的总人口约为10546人，其中，东兴村总人口为1884人，满族人口约占总人口的85%；东平村总人口为1072人，满族人口约占总人口的79%，这两村的满族人口占比较大。

人口状况：老龄化严重、空巢村现象严重、农业高级人才欠缺。

规划范围内的村屯常住人口老龄化比较严重，目前60岁以上人口占总人口的比例平均为22.9%，15年后老年人口将达到51.4%，有的村屯甚至高达54.4%。

村屯内18~45岁的人口基本全部外出打工，大部分时间在家种地的都是45~65岁的人口，空巢村现象比较严重，受过一定文化教育的年轻人口外流，缺乏具有现代农业知识的人才。

中国进入老龄化社会以来，呈现出老年人口基数大、增速快、高龄化、失能化、空巢化趋势明显的态势，再加上未富先老的国情和家庭小型化的结构叠加在一起，养老问题异常严峻。

Population scale: The total population of Xiangyang Town is about 10546, of which the total population of Dongxing Village is 1884, and the Manchu population accounts for 85% of the total population;the total population of Dongping Village is 1072, and the Manchu population accounts for 79% of the total population. The Manchu population of these two villages accounts for a large proportion.

Population situation: Serious aging, serious empty-nest village phenomenon, lack of senior agricultural talents.

The aging of the permanent population in villages and villages within the planning scope is quite serious. At present, the proportion of the population over 60 years old in the total population averages 22.9%. After 15 years, the elderly population will reach 51.4%. Some villages even reach 54.4%.

The 18-45-year-old population in the village basically went out to work. Most of the time people who planted their land at home was 45-65-year-old population. The phenomenon of empty-nest village was serious. The young people who had received certain cultural education were outflowing, and there was a lack of talents with modern agricultural knowledge.

Since China entered the aging society, there has been an obvious trend of large population base, rapid growth, aging, disability and empty nesting of the elderly. In addition, the situation of the old before the rich and the structure of family miniaturization are superimposed, which makes the problem of providing for the aged extremely serious.

下辖村屯名称	户数（户）	户籍人口数（人）	备注
东兴村	506	1884	满族人口约占总人口85%
东平村	292	1072	满族人口约占总人口79%
东胜村	625	2280	
石槽村	542	2053	
向阳村	474	1979	
东方红社区	641	1278	
总计	3080	10546	

年龄结构	人数
0~18岁	1748人
18~45岁	3380人
45~60岁	3003人
60岁及以上	2415人
总计	10546人

土地使用现状
LAND USE STATUS

向阳镇特色旅游镇规划范围涉及 5 个行政村和 1 个社区，总用地面积 50.1km²，其中，城乡居住用地面积约为 2.9km²，耕地总面积 53766.0 亩（1 亩 ≈ 667m²）。

土地使用状况：土地使用粗放、建设用地闲置严重、弃耕现象较多、土地使用不紧凑等。

土地使用粗放：由于村屯人口进入城市，导致许多土地被放置荒芜；另外，留在村屯种地的大多是老龄人口，耕作能力下降，导致耕种粗放。

建设用地闲置严重：建设用地闲置问题严重。一种是肆意圈占耕地改为建设用地；一种是征而未用或多征少用、早征迟用或征而不用。

弃耕现象较多：大项目建设征用农田的影响，导致农民错误地认为种粮食不如种树、种大棚（一般闲置），从而弃耕农田。

土地使用不紧凑：现状土地使用较零散，建设用地分布也较零散。耕地一家一户耕种，弃耕耕地与耕种用地混杂，不成规模；村镇建设用地零散分布各村屯，集中使用的仅有向阳工业小区。

The planning scope of Xiangyang characteristic tourism town covers five administrative villages and one community, with a total land area of 50.1 square kilometers. Among them, the construction land area of urban and rural residential areas is about 2.9 square kilometers, and the total cultivated land area is 53766.0 mu.

Land use status: Extensive land use, serious idle construction land, more abandoned farming, land use is not compact, etc.

Extensive land use: As the population of villages enters the city, many of the land is laid barren; in addition, most of the land left in villages is for the elderly population, which leads to the decline of farming capacity, resulting in extensive farming.

The problem of construction land idleness is serious: The problem of construction land idleness is serious. One is to reclaim the cultivated land and change it into construction land arbitrarily; the other is to expropriate but not more or less, sooner or later, or to expropriate but not to use.

Abandonment of farmland is more common: The impact of requisition of farmland for large-scale project construction has led farmers to mistakenly believe that planting grain is inferior to planting trees, planting greenhouses (generally idle), abandoning farmland.

Land use is not compact: The current situation of land use is scattered, and the distribution of construction land is scattered. Farmland is cultivated one by one, abandoned farmland and cultivated land are mixed, not large-scale; construction land scattered in villages and towns, concentrated use of only Xiangyang industrial district.

下辖村屯名称	总用地面积（km²）	耕地面积（亩）	宅基地面积（万 m²）	企业占地面积（万 m²）
东兴村	11.1	12941	57.9	10.6
东平村	5.1	4754	26.26	6.5
东胜村	4.4	10125	4.4	7.0
石槽村	8.9	7770	10.7	10.7
向阳村	10.7	8248	8.9	120.0
东方红社区	9.9	9928	27.0	—
合计	50.1	53766	135.16	154.8

收入与就业
INCOME AND EMPLOYMENT

收入状况：规划范围涉及的 5 个行政村和 1 个社区，总收入约为 1.18 亿元，平均人均收入 17000 元左右，略高于国家农民人均平均收入水平。这些收入来源主要为外出务工和农业种植业收入，每亩地大田仅收入约为 500~700 元，水稻、蔬菜略高一些。

就业情况：村屯 18~45 岁的人口基本全部外出打工，大部分时间在家种地的都是 45~65 岁的人口。

Income status: The total income of five administrative villages and one community covered by the plan is about 118.00 million yuan, and the average per capita income is about 17,000 yuan, slightly higher than the average per capita income level of farmers in the country. These sources of income are mainly from migrant workers and agricultural planting. The income per mu of land and field is only about 500-700 yuan, while the income of rice and vegetables is slightly higher.

Employment situation: The population aged 18-45 in villages basically go out to work, and most of the time people who farm at home are the population aged 45-65.

农业生产状况
AGRICULTURAL PRODUCTION STATUS

农业种植结构没有太大转变。主要种植玉米、大豆，部分种植水稻、蔬菜或大棚蔬菜，少量种植了果树等。徐家油坊承包地流转约 2600 亩土地种植大果榛子，唐文公流转了 900 亩土地种植大果榛子，另外在东方红农场还种植了一个大花草园，占地约 1200 亩。

农业种植规模也基本没有改变。基本是一家一户种植，整个规划范围内仅东兴村有五家种植合作社、东胜村有一家种植合作社、石槽村有一家种植合作社。

Agricultural planting structure has not changed much. Maize and soybean are mainly planted, some rice, vegetables or greenhouse vegetables are planted, and a small number of fruit trees are planted. Xujia Youfang contracted land transfers about 2600 mu of land to grow hazelnut, Tang Wengong transfers 900 mu of land to grow hazelnut, in addition, the East Red Farm also planted a large flower and grass garden, covering about 1200 mu.

Agricultural planting scale has also remained basically unchanged. It is basically a one-family planting. In the whole planning scope, there are only five planting cooperatives in Dongxing Village, one planting cooperative in Dongsheng Village and one planting cooperative in Shicao Village.

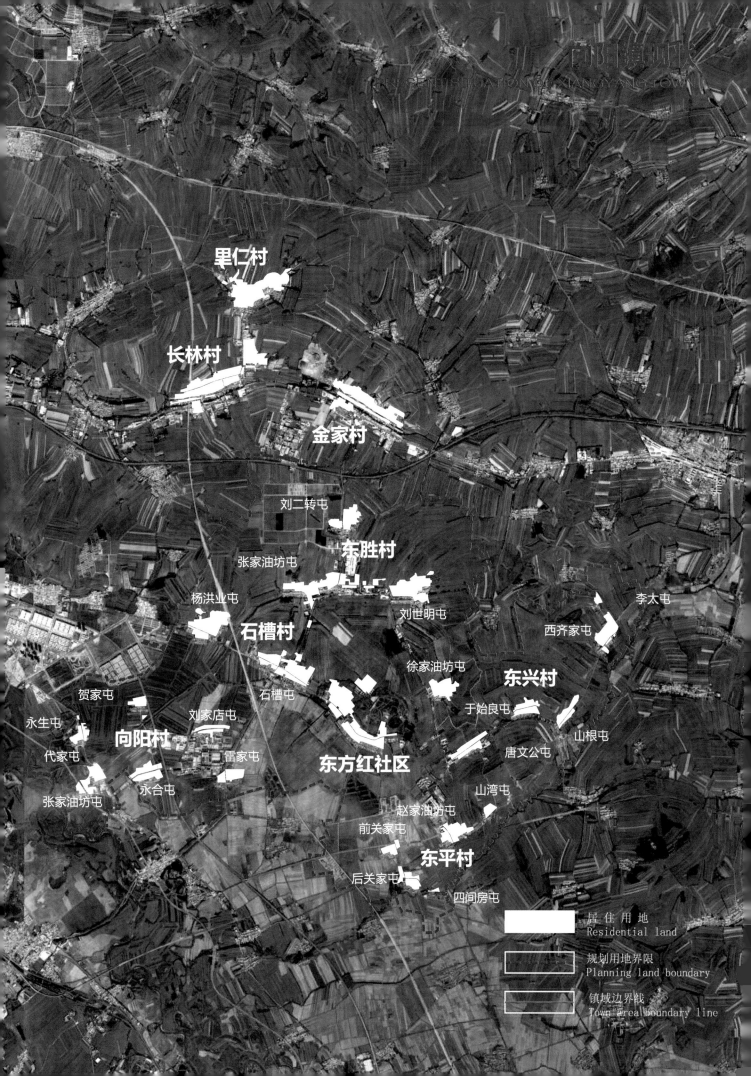

PUBLIC FACILITIES DISTRIBUTION OF XIANGYANG TOWN

畐仁村

长林村

金家村

刘二转屯

张家油坊屯

东胜村

杨洪业屯

刘世明屯

李太屯

石槽村

西齐家屯

徐家油坊屯

东兴村

贺家屯

石槽屯

永生屯

刘家店屯

于始良屯

山根屯

向阳村

代家屯

雷家屯

唐文公屯

东方红社区

永合屯

山湾屯

张家油坊屯

赵家油坊屯

前关家屯

东平村

后关家屯

四间房屯

居住用地
Residential land

规划用地界限
Planning land boundary

镇域边界线
Town area boundary line

了解从 1900 年到 2018 年哈尔滨地图和规划的演变。从过去的沿铁路发展到 21 世纪的常规环路结构。从"公司镇"到城市"蔓延"以及用途分区的规划。

Understanding Maps and Plans from 1900 to 2018. Evolution and current condition. From railway extensions to 21st century generic ring-roads structure. Plans from "company towns" to "extension" cities and to zoning by uses.

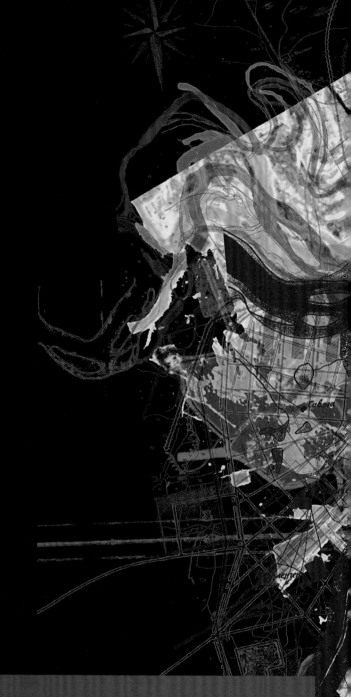

XIANGYANG IN HARBIN ALONG

THE HISTORY

历史演变

1966

2018

俯视跨越世纪的历史
THE HISTORY IN THE AERIAL VIEWS AT THE TURN OF THE CENTURY

哈尔滨市具有独树一帜的特定的形态特征和成长增长模式。它保留了其自历史起源以来所确立的城市景观结构，如规划布局及城市形态演变，让我们能够清晰地解读其演变与城市规划的发展，以及整个 20 世纪实施的城市规划。

哈尔滨是"因路而兴"的城市，是人类城市文明的奇迹。境内中东铁路连接哈尔滨、经长春，几乎直达北京，并经绥芬河跨境连接赤塔和海参崴。1898 年中东铁路的建设，将哈尔滨确定为中东铁路管理中心，揭开了哈尔滨城市建设的序幕。

中东铁路穿过哈尔滨城市核心区，一条跨过松花江向南、向西南方向驶向北京，主要顺应河流与其支流马家沟之间的自然平地。而另一条向东南方向驶向绥芬河，跨境达海参崴，这些是哈尔滨城市起源的历史痕迹。

在这里我们可以找到其与向阳镇现状相似之处。向阳镇域被哈牡高铁分割，江南中环路向北、向南轴向穿过，北侧的哈同高速公路、西侧的哈阿快速路（长江路延长线）是与城市连接的主要通道，这些足以促进城市向该区域的蔓延与扩张，给向阳镇带来了发展机会，但需明确限制粗放不经约束的蔓延，给定一个具体的解决方案。

The City of Harbin presents certain morphological characteristics and an urban growth model that makes it unique among the other Chinese cities. The conservation of the structures that defined the city landscape from its origin, like the product of planned layouts with project criteria, allows us to read its evolution alongside the developments of city plans, and the urban proposals that have taken place throughout the twentieth century.

The construction of the "Trans-Manchurian" train, with its branch directly linking Chita and Vladivostok diverging in Harbin, allows the almost simultaneous construction of the direct train to Pekin. At the original urban core of Harbin, the railway's route crosses the Songhua River to the south and bears right to the southwest toward Pekin, adapting to the natural platform between the major river and its tributary, Magiagou. With another turn to the southeast, a diverging train line goes towards Vladivostok, leaving three original branches, which signifies the first development of the city.

Here we can find a similarity with the current situation of Xiangyang. The area is delimited by the major bifurcation of the highspeed train and crossed by the axes of the territorial mobility that travels east. In the eastern section of the road belt, the G1001 expressway, becomes a principal connector that can facilitate the expansion of this area of the city, but also is an opportunity to establish a clear limit to the untamed expansion and favor a more specific solution.

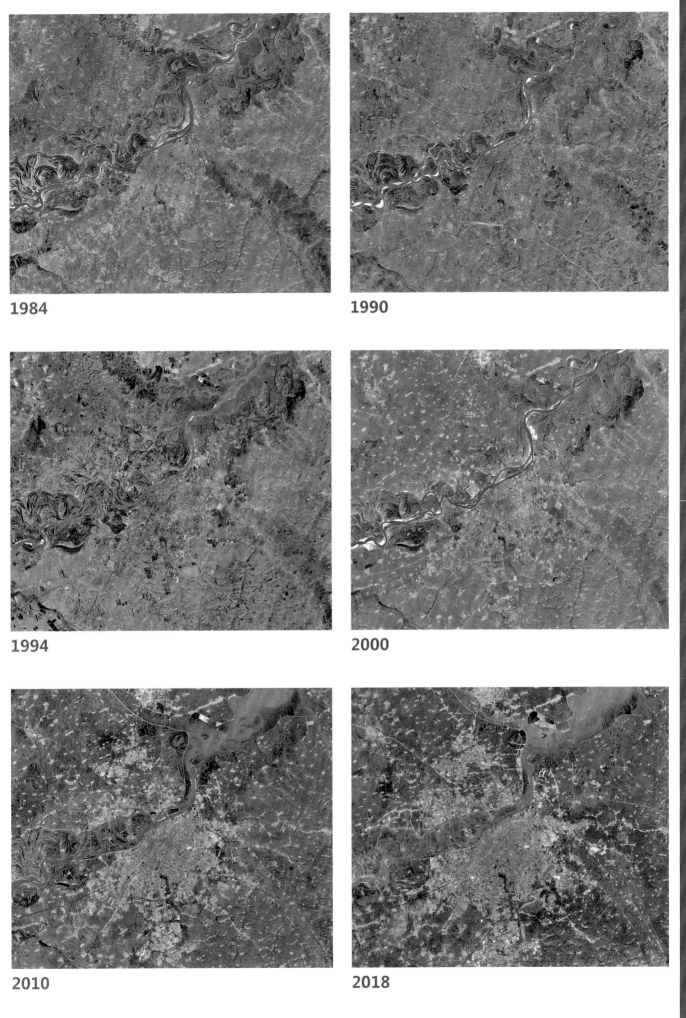

1984

1990

1994

2000

2010

2018

读地图解析向阳起源
ORIGIN IN PROJECTS OF INDEPENDENT IMPLEMENTATIONS

20 世纪哈尔滨的城市地图系列始于 1906 年由俄国绘制的城市地图。自从作为各种集市的交汇点以来，城市为其产业用途创建了有序的体系，并开始建设开发，以便为那些被城市活动吸引而来的员工及其家庭和其他人口提供服务和住房。

它由城市不同的独立片段组成，通过铁路网络形成公共的共享体系。秩序通常是精确的、几何的、非正交的，并且主干由新基础设施的轴线构成，具有标志性的城市结构完好地保存下来，形成了哈尔滨的"老城区"。

沿着河流，可以看到一系列不连续的活动以及它们如何分布于河滨。这些活动与铁路网络有着紧密的关系。

1907 年的城市地图（中文版）具有更简化的表示，可以更清晰地阅读之前地图中显示的信息。它们与后来发展了的城市的信息是相似的，从一张地图中人们可以看到对另一张的重新解读，然而 1907 年的地图给我们提供了更准确的地域范围，使我们更了解向阳地区，而且虽然没有到达该地区，但小而均匀分布的村庄，以及蜿蜒的被大量的树木所包围的阿什河的边界，形成了向阳区域聚落的聚集模式。

1907 年的第二张地图用俄语书写，表达了城市的制图细节，展示了建筑建设和基础设施，以及水文、地形、洪水区和散落在最接近城市发展第一圈外环整个地区的小村庄的情况。

前三张地图中的内容使我们能够解读出原始城市与铁道桥位于同一高度的平台上，并转向与河岸平行排布，产生城市最原始区域的"4"字形状特征。在这个平台上，可以保护城市免受可能的洪水泛滥伤害，河流边缘（与当今河流的边界重合）创造了一个中间缓冲区，这里是第一个大型火车站的开发区域所在地。

The series of the twentieth century city maps begins with the 1906 Russian map. Since its beginnings as a meeting point of various markets, many companies created ordered sections for its industrial use and the construction development for the service and housing of the employees, their families and the rest of the population that could be attracted to the city's activities.

It's made up of independent sections of the city that share the commonality of having physical contact with the railway network. The ordering normally is precise, geometric, and meant to be very orthogonal, with principal guidelines formed by the axes of the new infrastructure. The identities of urban structures have been conserved perfectly to create an "old city" of Harbin.

Together with the river, you can see a discontinuous series of implementations of activities that share a relationship with the railway network and how these tend to occupy the riverfront.

The 1907 Chinese map, with a more simplified representation, allows a clearer reading of the information presented in the previous map. They are similar in the information regarding the developed city, where one map can be referred as a reinterpretation of the other, although the map from 1907 gives us a better territorial scope that brings us closer to the area of Xiangyang and although does not reach the area, adheres to the territorial occupation model through small, homogeneously distributed villages, with the boundary of the Ashi River, which in its snaking qualities, appears surrounded by masses of trees.

A second map in 1907, written in Russian, represents the city in great cartographic detail, as much in the building constructions and the infrastructures, as in the representation of water, the topography, the flood zones, and the small villages disseminated throughout the area closest to the first ring of the city's development.

The content in these first three maps allows us to interpret that the original city is situated on a platform that reaches the railroad bridge and turns to align itself parallel to the riverbank. On this platform, protected from possible river flooding, and the edge of the river (which coincides with the border of the present-day river) creates an intermediate platform that is where the first developments of large train yards are located.

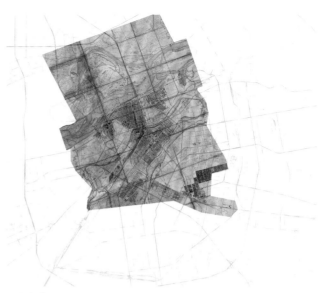

1906

1907

1907

城市扩张计划的城边地带
JOINS FROM THE URBAN EXTENSION PLANS

从 20 世纪 20 年代末持续到 20 世纪 50 年代，城市呈一种有秩序的增长，与城市规划及网络基本吻合，同时寻找潜力地区纳入城市中，并连接已开发地段。

规划图纸显示了大致完成的片段，有些情况下没有区分已建成建筑部分和预期扩展部分。道路网呈现了不同类型的道路：大道（更宽更具连续性）、林荫大道（中央有绿地空间）、街道……此外，从围合城市建设的街区边界的主要道路中，可以观察到城市形态增长的细节。

从基础设施的角度来看，这些规划为我们提供了有关铁路、道路以及在某些情况下河流利用的信息。因此，人们只能偶尔看到一个码头，没有其他信息可以确定其与铁路线的必要关系，这表明不同交通方式之间的距离很近。铁路与城市 20 世纪下半叶的城市地图反映出铁路线的变化（第一条传统的铁路线在 20 世纪中仍然存在），又出现了新的铁路线，同时期又有少量的铁路线消失，让位于更多其他的城市用途。

关于铁路的平面图证实，尽管有多条延伸线，铁路仍具有一定程度的持久性。保留下来的火车站仍保持其完整性，一直处于同一位置。

因此，我们的工作涉及的不仅是一个由铁路网形成的城市，更是一个必须将其网格融入城市布局的城市。不仅必须形态上相适，而且还要寻求解决方案，使城市不同碎片互相之间产生内在联系，具有连续性和关系，如基础设施的解决方案。关于如何整合片段的城市连续体，创造相关的流动性使区域更好地突破基础设施壁垒而互相联系起来，在这一点上哈尔滨市区与向阳镇具有相似性。

关于公路，在小村庄聚落之间相互密集连接的底图上，可以看出主要道路呈放射状，河流成为它们联系的最主要障碍。该路网界定了哈尔滨在地域以及经济方面作为桥头堡战略位置的重要性。放射状道路网由主干道构成，证实了它作为一种独特的道路结构持续了 20 世纪的大部分时间。

In this period that lasts from the end of the 1920s to the 1950s, the urban growth shows an order that responds to plans with urbanistic criteria put forth by urban extension (grid) models, that look for ways to integrate the new proposals into the preexisting city and to help join already developed pieces.

The maps show pieces somewhat completed, in some cases without distinguishing between the built section and the anticipated extension. The road network represents different types of roads: avenues (more width and continuity), boulevards (with central green spaces), streets...In addition, you can observe the growth detailed in the urban form, from the main roads that create the borders of the blocks where the city was constructed.

From the point of view of infrastructure, the plans offer us information about the railway, the roads and in some cases, the use of the river. Therefore, one can only occasionally see a dock, without other information that allows to determine the necessary relations with the railway that would suggest the proximity between different modes of transportation. In respect to the railway, the planimetric documentation for the second half of the century confirms the great seasonal inertia that supposes that occupations in this transportation system, given that new railway lines appear (in fact, the first conventional rail lines already existed at the middle of the century) are few in that in the same period have disappeared giving way to more urban uses.

The planimetric documentation regarding the railway confirms that the railway was meant to have some level of permanence despite receiving multiple extensions. The trainyard has remained intact and in the same position

Therefore, we are working with a city that is not only shaped by the railways network, but one that has had to integrate the network into its urban layout. It has had to act not only to adapt physically, but also to seek solutions that allow interconnections of different pieces of the city, looking for continuity and relationships, such as solutions with the infrastructure. This is the new parallelism with Xiangyang, in how to integrate into the urban continuum the relational flow that allows to connect this area with the city beyond the infrastructural barriers.

In respect to highways, on the base of the dense interconnections between small rural settlements, there is a radial scheme of major roads that has the river as its most major obstacle. This network demonstrates the importance of Harbin for its strategic location as a bridgehead, both in terms of territory and economy. The radial network of roads is made up of interior avenues, confirming that it is a unique road structure model that lasted throughout most of the twentieth century.

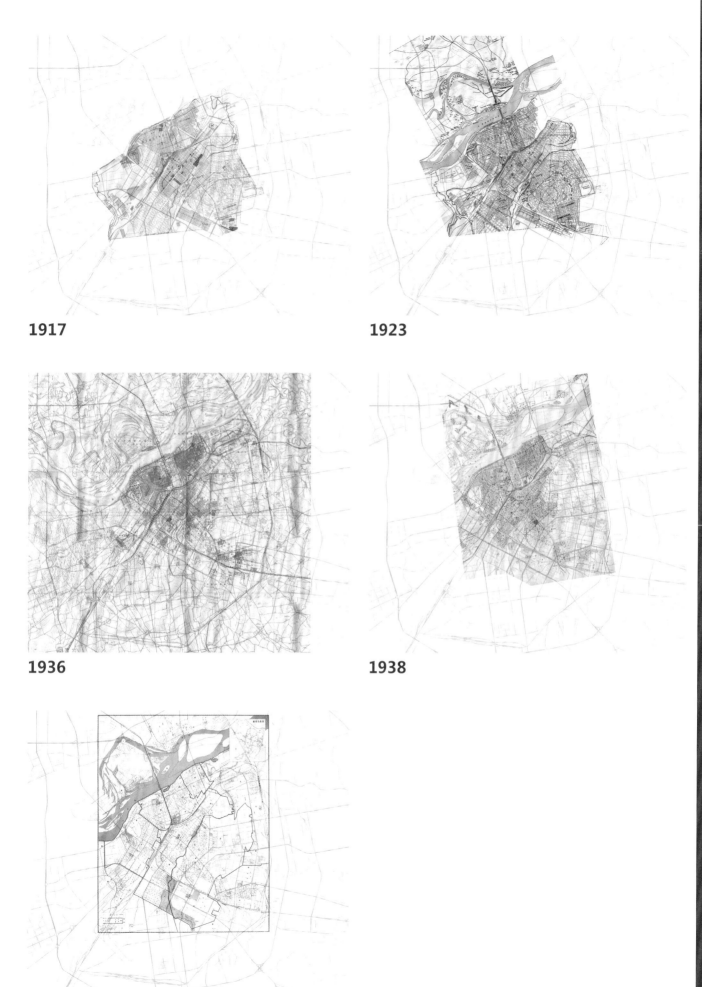

1917

1923

1936

1938

1946

无序增长的城市
UNRAVELING THE ORDERED GROWTH

20 世纪中叶开始，我们没有太多的信息，这一时期的规划信息很少。仅有 1964 年的规划图可以作为这一时期的参考，但这些信息突出了哈尔滨发展的非常有趣的一面。

它根据现有基础设施提供了一个非常清晰的地图：两个铁道桥确保了潜在扩张的经济可行性，但并未提供车行接口。在城市的南部和西部，城市领域范围贴着河的南侧，形成一个不规则的形状，在东部，则显现出一条限制城市增长的铁路线。跨越这条线则是一个对向阳镇域非常重要的铁路网的开端。

当我们将该地图与当前地图进行比较时，在 20 世纪 70 年代至 2000 年之间建造的交通大干线主导了城市肌理组织。这些为了快速通达而创建的元素，通过各种随机的发展逻辑，产生新的城市扩张蔓延，而这些逻辑几乎没有与先前存在的城市肌理结构相结合。它来自一种分层结构模型，并未考虑区域范围，只通过一些通道或转换节点的简单接合相连。

这些城市增长体系可以与特定的群体类型相关联，这些非常清晰的城市增长体系（城市肌理演变和道路体系），尽管没有理由将这两种模型（道路和体量）联系起来，但它们始终保持一致，并且至少那时对已建成城市的根本转变具有并行的效果。它由包含巨型住宅结构的街区建筑组成，形式上非私人化、同质化，不会创造出有趣的城市空间。至此，一个源于非常有条理和有凝聚力的规划，以有秩序的增长体系而闻名的城市，忘记了它先前自身的范例，而采取了恣意的城市蔓延形式。

人们可以理解 1950 年以前的开发，对于当前城市的负面影响：南部和西部的扩增区域，以及河流另一侧的发展方式（归功于新的桥梁）创建了大型住宅区的模式。在东部区域，铁路线上已经建立了一个大型工业区，以致至今在这个方向上仍限制了城市扩张，这使得向阳镇域的大规模区域保留完好。

From this period, we do not have too much information as the city's detailed mapping techniques were replaced by systems resembling zoning plan maps. Only the 1964 plan can be useful as a reference for this period, but the information highlights very interesting aspects to the city's development.

It results in a very clear map in regards to existing infrastructures: two railway bridges that ensures economic viability of a potential expansion, but do not offer access to motor vehicles. In the southern and western side of the city the urban occupation sticks exclusively to the south side of the river to create an irregular form and to the east, it shows that it has a rail line confining the city's growth. Beyond this line appears to be the beginning of a network of railyards that will be very important for the sector of Xiangyang.

When we compare that map with the current one, the large arteries constructed between the 1970s until 2000, dominate the city's urban tissue. These elements, created for quick access, favor the development of new growths through various random logics, barely structured with the existing urban fabric. It comes from a hierarchical structure model that does not take the territory into account except to look for easier points of passage or exchange nodes.

These urban growth systems can be associated with a particular massing typology. These very distinct urban growth systems (urban tissue development and road systems) that, although do not have a reason to associate the two models (road and massing), they coincide with the time and have a parallel effect of radical alterations of the developed city until then. It is made up of the block constructions containing mammoth residential structures, formally impersonal, homogenous, that do not create urban spaces of interest. Suddenly, a city that was known for very orderly growth systems that were derived from very structured and cohesive plans, forgets its precedents and adopts an indiscriminate growth.

Despite the negative aspects of the current city, one can appreciate what had been developed pre-1950: the areas of the growth in the south and west, as well as how the development at the other side of the river (thanks to the new bridges) exclusively creates the model for the massive residential complexes. In the east, the rail line has established a large industrial zone that has restricted the urban expansion even today this same direction, which has led to the conservation of the largely unaltered Xiangyang.

1954

城市规划的演变
EVOLUTION OF URBAN PLANNING

　　前一章节已阐述了哈尔滨扩张中一些重要的城市的转变，本节关注于可以解释这些转变的城市因素，但并不是对城市规划的宏观论证，而是找到在当前城市中永久保留下来的规划的各个方面。这是为了列出影响这一过程的机制。城市发展较好地遵循了这些规划，并转为完全可识别出的城市实际状态。

　　我们研究的第一个规划是1902年的俄罗斯编制的规划，这与铁路的建设相吻合，这使我们相信当时已经达成协议，这一时期，铁路成为与松花江交汇点一样的突出位置。该规划优先考虑指定用于铁路建设的区域和在其轴上创建的城市碎片。1906年和1907年的城市地图可以很好地反映1902年的规划，因为它绘制了一个完美的由铁路和代表预期区域的城市发展区勾勒的范围。

　　二十年后，从1917年和1923年的地图可以看出，城市超越了之前的划界，呈现出相互关联的几何形式，并有大面积的扩张。铁路网嵌入空间之中，从而形成了城市逐步向外的增长，因此也形成了被保留在当前城市中的城市碎片。

　　20世纪30年代后期，1936年和1939年各自的规划（或同一地图的两个版本）给出了新的补充，虽然未能实现某些重

1902

In this section, we will look briefly at the urbanistic factors that can explain some of the most important transformations in the growth of Harbin, which we have laid out in the previous section. It is not as much to create a grand argument about the planning of the city, but instead to find aspects of the plans that have stayed permanent in the current city. This is to list the mechanisms that have influenced this process. The development has stayed faithful to many of these plans, which has translated to completely recognizable sections of the city.

The first plan that we study is the 1902 Russian plan, that coincides with the construction of the railway, which leads us to believe that an agreement was made to adopt the various initiatives that gave way to new infrastructure in a point as singular as the crossing of the Songhua River. This plan prioritizes areas designated for railway construction and urban pieces that are created on their axes. The city that showed the urban maps of 1906 and 1907 are well reflected in this plan of 1902, as it draws a scope perfectly outlined by both the railroad and the urban development representing a preestablished area.

Twenty years later, the 1917 and 1923 plans went beyond the previous strict delimitation and presented interconnected geometric forms, with large areas of expansion. However, they portray inserted space between the railway networks, creating a progressive outward growth. It was partially developed in pieces that have been conserved in the current city.

In the late 1930s, the 1936 and 1939 plans(or two versions of the same plan), give new additions that, although failed to materialize certain important aspects, included these significant facets:

– A proposal for two bridges with major urban roads, along with the two railroads that exist today. This permitted the expansion to the other side of the river. However, this did not materialize.

1917

1923

要方面，但涵盖了以下重点：

①关于主要城市道路的两座桥梁，以及现今仍保留的两条铁路的提案。这保证了城市将扩展到江的另一边。但是，这没有实现。

②细胞组织状的城市网格结构，由环形主干道围合，并沿着火车线向东。后来的提案（包括当前的）一直在继续发展这一想法。

③提出同心圆式的城市增长，将环形公路和放射状公路转变为主干道。这个提案部分实现了。

④一项工业用途的北部和东部城市的扩建计划。经过后来的提议，这些工业区在中心城区和向阳镇之间的中间区域发展起来。

对于其他方面来说，城市扩张也将建成区考虑在内，这在总规中有详细说明。一些通往高速公路的街道被重新塑造成与部分环形道路相连的主要道路。定性地说，20 世纪 20 年代的规划图表达了由建成区的碎片扩张转向国际都市的增长。

20 世纪中叶以后，1953 年的规划图已呈现出根据城市分区方法的制图。它是一种非常简单且抽象的图面表达方法。从中可以看到我们之前已经提过的城市模型的基本网络，但它更多地参考了土地用途，而不是具体功能方面。规划着重表达了两方面：连接公园的城市通廊和铁路线。然而，尽管表达了诸多铁路线的细节，铁路的发展在这一时期并未起到非常重要的作用。

– A structuring of the urban network as a tissue, surrounded by main roads that establish ring roads following the train line to the east. Later proposals (including current) have continued to develop this idea.

– A proposal for a concentric growth, converting the ring road and the radial roads, to principal arterial roads. This partially materialized.

– A scheme that would facilitate possible northern and eastern urban expansions for industrial uses. After later proposals, these industrial areas developed in the interposed zone between the center and Xiangyang.

For the rest, the growth continues to take the developed city into account, which is addressed in detail within the general proposal. Some streets with access to the highways became reformulated into major avenues that interconnect with partial ring roads. Qualitatively, the 1920s maps created the growth beyond the expansion of the pieces of the developed city to a global proposal.

After the middle of the twentieth century, the 1953 plan creates its cartography based on the city's zoning methods. It is a very simplistic and arguably abstract graphic expression. One can recognize the basic traces of the urban model that we have previously addressed, but in reference more towards the uses than the functional aspects. The very basic scheme of the map mainly emphasizes only two aspects: urban corridors linking parks and rail lines. However, while the railway is presented in great cartographic detail, railway development did not have a formative role in this era.

1936

1939

1956

1981 年的城市规划给我们提供了两个新的重要信息。一方面，该规划配置了更宽的网格规模，以纳入一个高容量系统，包含三个部分：

①阿什河上建造的 202 国道桥梁（虽然略有位移）。

② 20 世纪 80 年代开始建造的 G10。

③开始建造放射状的高速公路，规划有表达但无法确认，但可以假定规划已经考虑了城市发展的连续性。

另一方面，要参考东部工业区的规划，才可以理解城市最终将朝着向阳镇区域发展，因为它标记了城市朝着这个方向进一步发展的障碍。只有通过这个区域的道路，才能结束连续性的增长。

在接下来的十年中，道路的发展与新城市结构的规划同时进行，这种结构为新的增长整合创造了空间区域。1996 年，道路网络已完全由已经执行规划的区域或正在进行的项目确定，该规划已经勾勒出沿着主要道路走廊的增长轴，同样也指向东部。2011 年的规划显示了已经开发的道路的总体情况，尽管周边地区仅显示现有主要道路的延伸。城市向南、向西扩展，特别是向北穿越河流扩展。东部，城市边界仍停留在工业区和阿什河。

在过去十年中，道路建设已成为改变城市早期增长模式的决定性因素。值得注意的是，可以看出划定 20 世纪 30 年代后期的城市发展第一圈环带的网格形态特征，是如何被随后几年的城市扩张完全分割开的。

The 1981 plan presents two new important aspects for our proposal. On one side, the plans configure a broader network scale to incorporate a high capacity system with three main components:

a. The bridge over the National 202 River was constructed (although slightly displaced).

b. The beginning of the various G10 constructions in the 1980s.

c. The beginning of some of the future radial highways that, if the format of the available map does not confirm it, one can assume that the map already portrays the continuity.

On the other hand, to understand the city's eventual development to Xiangyang, one must refer to the plan for the large Dongguangcun industrial area, which marked the barrier for further urban development in that direction. Only the road axes that cross this sector would end up generating a growth with some level of continuity.

In the next decade, roads developed in tandem with plans for new structures incorporating and creating areas for new growths. In 1996, the road network was completely defined by the already executed districts or by the projects in progress, and the plan already outlines axes of growth following the principal road corridors, as well as to the east. The 2011 plan shows the totality of the already developed roads, although the peripheral territory only appears occupied by the extension of the existing major roads. The city expands towards the south, west and especially to the north crossing the river. On the east, the urban border continues in the industrial district and the Ashi River.

In the last ten years, the highway construction has been a determining factor to change the urban growth model of the early periods. It is significant to observe how the morphological character of networks that defines the city's development first ring in the late 1930s becomes completely unraveled by the expansionism of the following years.

1981

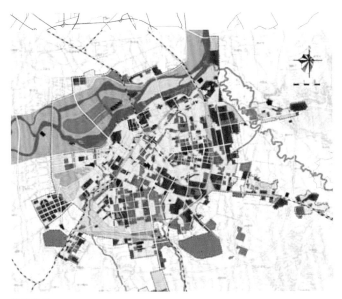

1996

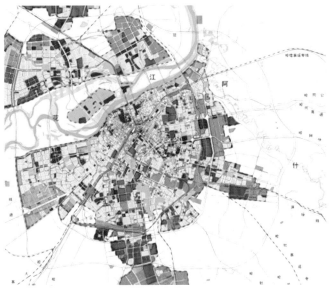

2011

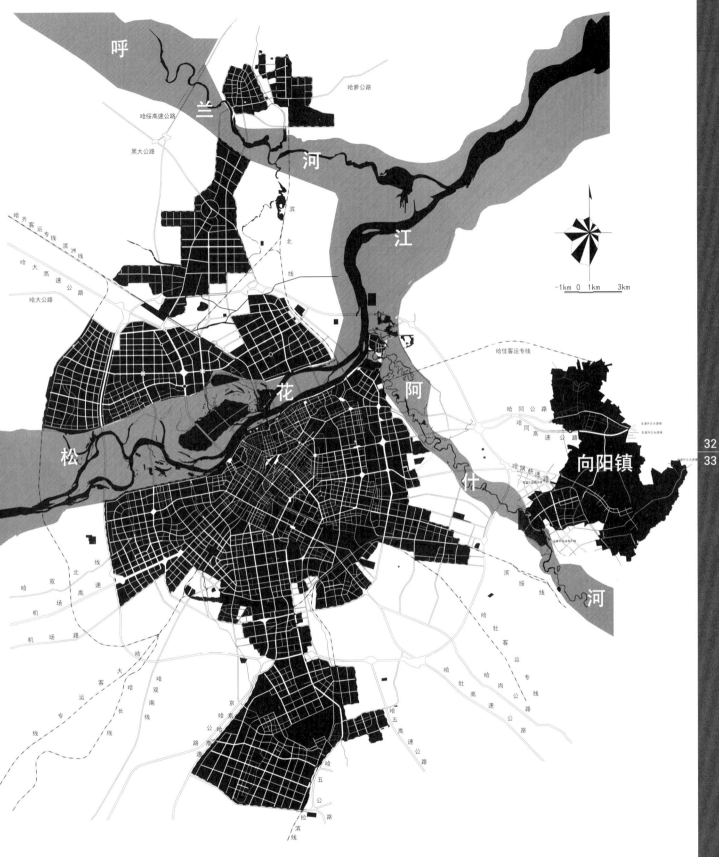

呼

兰

河

江

松

花

阿

什

河

向阳镇

哈蒡公路

哈绥高速公路

黑大公路

哈齐客运专线 滨洲线

哈大高速公路

哈大公路

滨北线

哈住客运专线

哈同公路

哈同高速公路

哈阿快速路

北线

哈双高速

哈机场线

机场路

哈客运专线

哈大客运专线

哈双南线

哈长线

京哈公路

哈泉哈高速

哈五公路

拉滨线

哈牡客运专线

哈牡高速公路

哈尚公路

哈五公路

滨绥线

-1km 0 1km 3km

城市肌底认识

RAIL INFRASTRUCTURE

AND THEIR CONNECTIONS TO

BOAT DISTRIBUTION

CREATING A POWERFUL

NETWORK

基础设施的演变：不同年代的铁路
INFRASTRUCTURAL EVOLUTION: RAIL, DIFFERENT HIERARCHIES

— 1901
— 1906
— 1936
— 1956
— 1992
— 2011
— 2017

哈尔滨的演变可以被每座铁路桥梁的年龄所解释
HARBIN'S EVOLUTION CAN BE EXPLAINED BY THE AGE OF EACH BRIDGE

铁路与安全的自然河岸
RAILWAY AND THE SAFE NATURAL PLATFORM

铁路基础设施是这个城市最具特色的元素之一，因为铁路环线比高速公路的主干网出现得早得多，早在 20 世纪 80 年代和 90 年代城市有效地扩展到这个领域之前，河流上方的铁路桥就先于道路出现，向北的工业扩张也在这一基础设施的基础上发生。

20 世纪初的铁路发展标出了哈尔滨城市最初飞地的位置。最初，它源自松花江南部桥头的交易商品的台地，该台地使用了不受河流周期性洪水影响的自然平台。从这一点开始，连接北京和海参崴的这两条铁路走廊将这一城市飞地带入高光时代。

20 世纪 30 年代以后，城市开始向东开发一条火车路线，将城市连续体封闭在一条完美的环带中，环带与河流北部相接并支持不同的火车站，作为内城主要工业发展的起源点。直到最近 21 世纪，高速铁路的建设才为哈尔滨铁路系统的发展带来新的变化。

The railway infrastructure is one of the most characteristic elements of the city, given that the ring of railways arrived much earlier than the arterial network of highways: the railway bridges over the river preceded the roads and the industrial expansion to the north appears at the base of this infrastructure, much before the city effectively expanded to this territory in the 1980s and 1990s.

The early twentieth century railway development marks the location of the original urban enclave of Harbin. The area developed from an interchange platform for exchanges of goods at the meridional bridgehead of the Songhua River, that uses a natural platform safe from the periodic floodways of the river. From this point, the rail corridors towards Pekin and Vladivostok highlight the urban enclave.

After the 1930s, the city started to develop a train route to the east that enclosed the urban continuum in a perfect ring that meets the north of the river and supports different train platforms, as the point of origin of the principal industrial development in the inner city. It would not be until recently in the twenty first century, that the construction of the highspeed train would create a new change in the development of the Harbin train system.

哈尔滨可以从水系、地形和铁轨等方面被解读
HARBIN CAN BE EXPLAINED BY THE WATER, TOPOGRAPHY AND RAILWAY

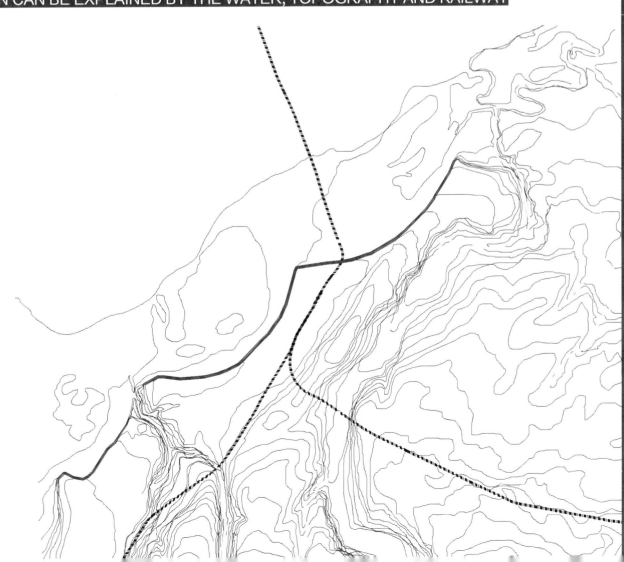

令人印象深刻的人类基础设施在很多情况下塑造了流动的自然渠干，除此之外，历史图纸向我们展示了松花江从偶尔周期性的洪水通道发展成为非常宽的通道，城市中其余的河渠汇入其主要通道。这条河流汇聚了哈尔滨的几种水文形态，这对我们的研究是很重要的。举以下例子：

　　①马家沟——处于中心城市连续体的水系，形成了哈尔滨原始平原南部的自然界限；

　　②阿什河，这是向阳镇的边界，现今处于一种自然状态。

　　松花江大坝体系的建设可以很好地控制主要水流。东北方市郊的大坝为滨河平台提供更多的安全保障，而西南方的大坝为市民休闲活动提供滨水空间。

　　然而，这座大坝杜绝了哈尔滨大面积商业区与俄罗斯萨哈林岛（太阳岛）江岸之间有交通联系的可能，这种交通联系是在二十世纪的大部分时间内逐渐建立的，它意味着松花江曾经可以通航。由于这些原因，哈尔滨被创建为一个商业窗口，从东北方的边缘到国家的中心，甚至通过西伯利亚铁路通向欧洲。

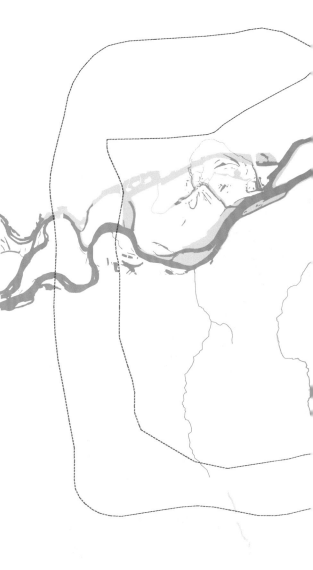

Beyond the current reality, in which the impressive human infrastructures have molded in many cases the natural channels of the flow, the historical cartography shows us how the Songhua had an occasionally very large channel from the periodic floodways, that leaves the rest of the changing meanders that complement the principal channel. This river sources several water forms that flow through Harbin, which is what matters most to our research, we can mention:

a. the Magiagou River—very internalized in the central urban continuum and that forms the natural limit in the south of the original platform of Harbin;

b. The Ashi River, which is the border of the project area and is today in a generally natural state.

A system of dams allows for superior control of main water flows. The dams in the northeast metropolitan area provide safety for the riverfront platforms, while the southwest dams maintain water for civic and leisure activities.

This dam nevertheless makes it impossible to have a transportation function that during much of the twentieth century had been established between the large area of commercial exchange for Harbin and the coast, in front of the island of Sajalin, showing that the Songhua was once navigable. For these reasons, Harbin acted as a commercial gateway from the edge of northeast to reach the center of the country, and also towards Europe with the Tran Siberian.

河流和溪流：松花江的控制区

RIVERS AND STREAMS: THE CONTROL ZONE OF SONGHUA RIVER

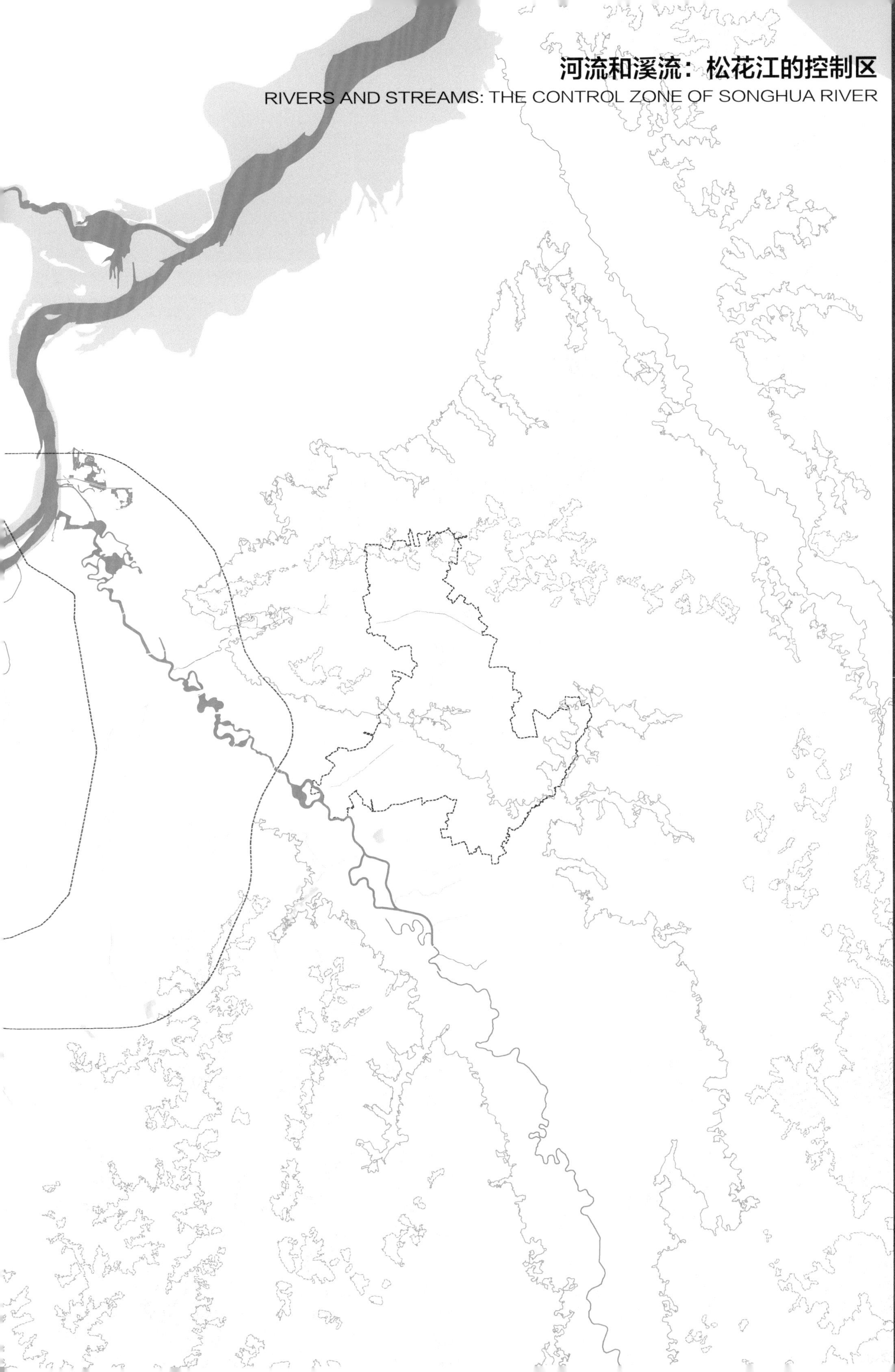

城市用地演变
LOGISTIC SURFACE LOCATION AND ITS EVOLUTION

铁路桥与城市用地
RAILWAY BRIDGES, LOGISTICAL PLATFORM STRUCTURE

铁路开发中的两个首次大的"跳跃"与松花江上前文提及的桥梁有关：

· 1901 年原始的中东铁路图纸显示了一座铁路桥，但由于其他更大容量的类似基础设施的建设现在已被废弃；

· 1936 年的图纸显示已经建造了跨松花江的铁路大桥，这使得主线拥堵减少，并为江的另一边提供了服务。

城市用地的起源可以被认为是铁路结构的自身演变。第一座桥通往太平街平台，首次实现了南线和东线分流的宏伟方案。它还在铁路和河流之间创造了一个空间，很快被各公司用来创建一个易于到达的区域来发展他们的项目。这个初始模型的饱和填充了其他的内部飞地，这些飞地在 20 世纪 20 年代之前持续地变得更具吸引力。东向新线路和新桥梁的建设，为中心区域的外围地带提供了新的铁路平台空间，这也使得质的跳跃主动发生。

The two first large "jumps" in the railway development involve the aforementioned bridges over the Songhua River :

· The original 1901 maps of the Tran Siberian line display a railway bridge that is now in disuse due to a similar infrastructure of greater capacity.

· The 1936 maps already show the constructed E Binjiand Bridge, that would allow the decongestion of the principal line to offer services to the other side of the river.

The origin of the city's logistical implementations can be thought of as its own evolution of the railway structure. The creation of the first bridge that accesses the natural platform of Taiping Street, supposes the first grand implementation that leads to the principal bifurcation between the lines towards the south and east. However, in addition, it creates a space between the railway and the river that was soon used by various companies to create an easily accessible area to create their projects. The saturation of this initial model complements the other more interior enclaves that continue to become more attractive until the 1920s from the principal tracings. That should have guided the initiative to a qualitative jump, with the creation of the new line to the east and the new bridge that would give a place for new rail platforms in peripheral terrains in the central area.

世界上再没有其他的城市拥有如此之多的大型桥梁
THERE IS NO OTHER CITY WITH SO MANY AND LARGE BRIDGES

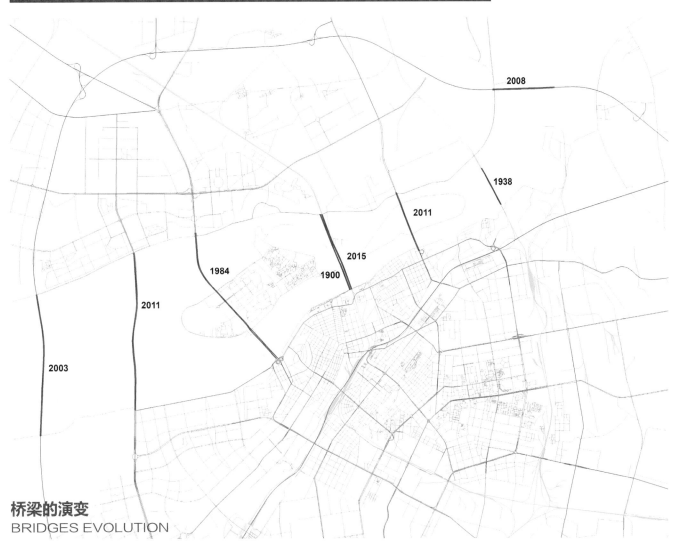

桥梁的演变
BRIDGES EVOLUTION

主要层面的叠加：地形、水文、聚居点、基础设施……
SUPERPOSITIONS OF PRINCIPAL LAYERS: TOPOGRAPHY, WATER, SETTLEMENTS, INFRASTRUCTURES...

像任何城市的结构一样，哈尔滨是形态特征、功能和连接彼此不同方式的关系等各个层面的叠加。地形、水文、聚落、功能和基础设施是始终呈现的基本要素，但是以不同的方式解读，相比之下更具相关性。

基地是一个地形系统，如水对地域的侵蚀以及由此形成的地形之间的相互作用，构成了城市形态。平台、河谷、渠化运河、高地和斜坡这些可以作为公共空间或因基础设施而形成的地形都包括在内。这些都遵循它们自身的逻辑，即如果考虑到用地，其功能性将更多地与城市功能，或活动和住宅区的关注点，以及它们与其他城市系统之间相互关系的必要性有关。

另一方面，村庄区域的传统做法值得一提。人们可以注意到各种规律。例如村庄的大小，平原地区的较大而高地地区的较小；或者村庄之间的距离，虽然总在减少，但与人口的规模成反比；或者是所有相邻的村庄之间呈三角形的相互连接的路网。然而，具有更重要的特殊性的是，我们可以看到双模式的存在：一种是由一组同质的节点组成，另一种是在一些谷地中形成了线性主干及其支线的树枝模式。

不同类型用地的扩张，结合创造了一个主轴体系，它与哈尔滨的放射状的谷地相关联，与结构上侧向的复合区相关联，与在它们之间起到区域连续作用的城市布局相关联。

Like any city's structuration, Harbin's form is the superposition of various layers of morphological character, functional and of relations that are different from each other in the way that they connect to each other. Topography, water, settlements, uses and infrastructures are foundational elements that are always presented, but are interpreted in distinct ways, giving more relevance to some over others.

At the base is an orographic system, like the interplay between the water's erosion on the territory, and the resulting topography that outlines the urban form. These include platforms, river valleys, channelized canals, elevations and slopes that can be incorporated as public spaces or that are formed by infrastructures. These follow their own logic that, if the territory is taken into consideration, the functional nature would have more to do with the urban functions, or the focus of activity and residential areas and necessities of interrelation between them all and other urban systems.

On the other hand, the traditional implementations of village areas deserve mention. One can note various tendencies like the sizes of the village, such as the larger ones in the flatlands and the smaller ones in elevated areas; or the distances between the villages, always reduced, but inversely proportional to the size of the populations; or the network of pathways that triangularly interrelate all the villages between the adjacent ones. However, of more significant particularity, we can mention the existent dual model—formed on one side by a homogenous set of nodes, and the other by the dendritic model that create in some valleys the linear occupations and its tributary streams.

The combination of different types of territorial expansion creates a system of principal axes, associated with the radial valleys of Harbin and with lateral complexes that structurally hold those, related with an urban layout that fills the territorial continuity between them.

哈尔滨周围有着大量的农业聚居点
LOTS OF AGRICULTURAL SETTLEMENTS ARE AROUND HARBIN

与基础设施相关的增长模型
GROWTH MODELS ASSOCIATED WITH THE INFRASTRUCTURES

通过前文对城市肌底的认识，可以理解哈尔滨城市增长是伴随着共同而必要的基础设施的发展的，可分为三种模式：

①城市起源：铁路占主导地位。1900—1910 年，虽然它不是城市结构的一个要素，但它让不同的区域因直接联系而不再独立。

②后期演变：城市紧凑发展，通过大型城市轴线相互连接。铁路分流的现状需要同心圆式的发展方案。铁路塑造了城市形态。

③现状解读：放射状高速公路，引导空间地域呈随机的形式发展，几乎没有尊重先前存在的肌底形式。它将大型建筑作为一个建筑单元，将建筑物与街道的联系断开。功能可达性的重要元素在它所创造的扩张中完全无法实现。

1900—1910: 铁路占主导地位
1900–1910: The railway dominates

哈尔滨已经拥有了非常强大的基建系统
HARBIN ALREADY HAS A VERY POWERFUL INFRASTRUCTURE SYSTEM

As a synthesis of the elements that we have considered up until this point, we can understand that the city's growth and the corresponding and necessary infrastructure development, respond to three models:

a. 1900–1910: The railway dominates. Although it is not an element of the urban structuration, the necessity of direct relationships between different districts generates independent areas.

b. Later evolution: The city grows compactly, with interconnected forms through large urban axes. The presence of the railroad as a bypass requires concentric solutions. The railway creates the urban form.

c. Highways: the territorial occupation opens itself to the random form and with little respect to the preexisting forms. It incorporates the mass building as a unit of construction that tends to disconnect the building from the street. The significant elements of functional accessibility are morphologically foreign to the growth that it creates.

后期的演变：城市紧凑发展
Later evolution: The city grows compactly

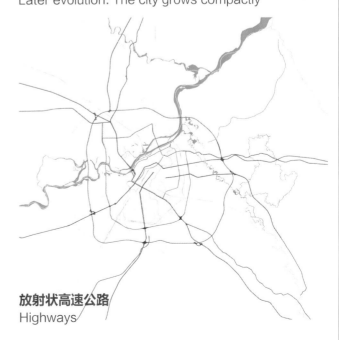

放射状高速公路
Highways

以上分析，帮助我们了解向阳镇在哈尔滨城市扩张中所处的位置。

向阳镇位于哈尔滨东部，处于哈尔滨的第二层城市增长环内。根据对哈尔滨城市与乡村布局模式的研究，即基于等距的中心和受控的大小规模而形成的如今周边乡村盛行的风格，可以发现向阳镇保留了乡村聚落的氛围。

然而，由于其具备靠近市中心的便利位置，这将导致哈尔滨诸多大型基础设施越过该区域，就像第一层城市圈通过城市的连接体而形成城市化的集聚一样。最新的城市扩张计划明确了这个认识，即扩大这种中心化城市的模式。

对哈尔滨20世纪的发展及历史与目前发展规划的分析，为我们对向阳镇的研究提供了指导方针，对新增长城市碎片评定提供了评价标准，对潜在于当今城市中的不同模式研究提供了分析基础。

这些帮助我们理解如何打造向阳镇的增长模式，这将是最充分有利的导则，以保证这种平衡城市必要性扩张模式的顺利实施。同时，提出实施解决方案，让人们充分理解向阳镇区域是城市联动和可持续的场所，并重视其最独特的地理特征。

The location of Xiangyang to the east of Harbin, in the city's second expansion ring, has preserved the colonial ambience according to the rural model of the city layout, based on equidistant centers and with the controlled size that today upholds the nearby rustic land.

However, the proximity to the city center has caused the large infrastructures that converge in Harbin to cross the area like the first territorial occupation for a future urbanization and absorption through the urban continuum. This expansion that is listed in the most recent plans confirms a new focus to expand the model of the centralized city.

The analysis of how the development of Harbin in the twentieth century, with both the historical and current development plans:

a. gives us the guidelines,

b. the criteria to join new urban pieces,

c. the different underlying models in today's city

This allows us to understand how:

a. to create the growth until Xiangyang and which would be the most adequate guidelines to implement a model that balances the urban necessities of growth.

b. the implementation of solutions of opportunity that allow an understanding of this area as a space of urban linkage and sustainability in the occupation of the territories valuing its most unique geographic characteristics.

规划与现实之间曾有着良好的关系，但是现在的问题需要被解决
THERE WAS A GOOD RELATIONSHIP BETWEEN PROPOSED AND BUILT, BUT GAPS MUST BE SOLVED

规划总图 2011—2020
Plan 2011-2020

规划道路 2011—2020
Road plan 2011-2020

规划总图（修编稿）2017—2020
Plan 2017-2020

呼兰区方岭镇

呼兰区乐业镇

宾县糖坊镇

呼
兰
河

巨源镇

刘业镇

温馨城

民主镇

永源镇

方宝镇

江

阿

向阳镇

万亩松江湿地

花

什

团结镇

松

河

成高子镇

阿城区料甸镇

新发镇

幸福镇

黎明镇

阿城区新华镇

王岗镇

朝阳镇

榆树镇

运

平房镇

土红旗乡

粮

双城区五家镇

河

双城区周家镇

CITY AND AGRICULTURAL PATTERNS

主城区与向阳镇

哈尔滨地域结构呈现三个维度：

①哈尔滨主城区；

②周边小村落；

③均匀散落在农业区域中的小型偏远乡村聚落。

这些由保证了可达性的铁路所连接的城市外围中心点为这个城市和河谷附近其他城区提供了同质特征的"马赛克"。

从历史上看，哈尔滨的道路呈放射状，这些通道向附近小的村庄延伸，或者向更远的城市延伸。它们为放射状的高速公路网奠定了基础，在许多情况下都是沿着作为支流廊道的自然谷地。

在景观方面，要继续聚焦基本结构，这与要研究的周边地区的农业特征有最密切的联系。它来自不连续聚落之间的步道网络，这些聚落根据不同的布局和地形地貌与地块结构耦合。未来的规划方案必须考虑到这点。

We can say that the occupation territory structure presents three levels:

a. Harbin as a center.

b. Surrounding small towns.

c. Small rural settlements uniformly dispersed through the agricultural territory.

The sum of these peripheral centers linked by the trails that guarantee accessibility, offers a mosaic that is homogenously characteristic of this city and of others nearby in the river valleys.

Historically, Harbin presents a radial network of trails that extend towards the near small villages or beyond that area towards further cities. They have configured the basis for the radial road network, following the natural valleys as corridors for communication in many cases.

In terms of landscape, we need to continue focusing on the foundational structure, the most intimate tie to the agricultural character of the peripheral region in which our study is situated. It comes from a network of small paths between the discontinuous settlements that couple with the parcel structure following the different layouts and topography of the territory. Future proposals have had to take this into account.

聚落形态
SETTLEMENT PATTERNS

3公里网格
3km GRID

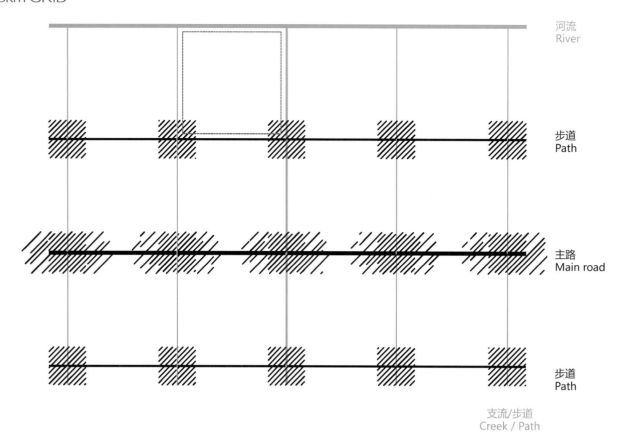

河流
River

步道
Path

主路
Main road

步道
Path

支流/步道
Creek / Path

1公里树突状节点
1km DENDRITIC – ARBOREAL NODES

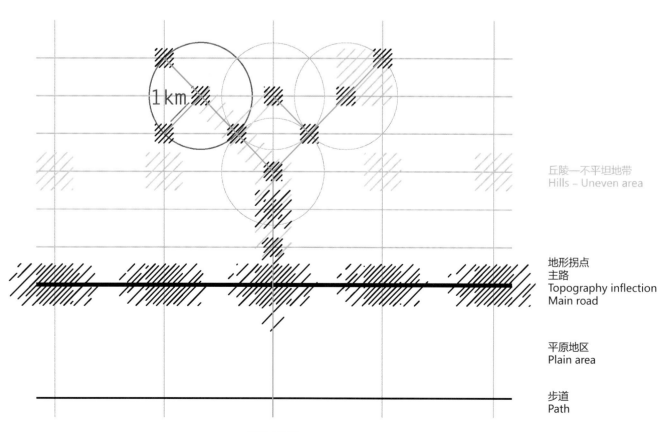

1 km

丘陵—不平坦地带
Hills – Uneven area

地形拐点
主路
Topography inflection
Main road

平原地区
Plain area

步道
Path

河谷 / 步道
Valley Basin / Path

常规平原
PLAIN – REGULAR

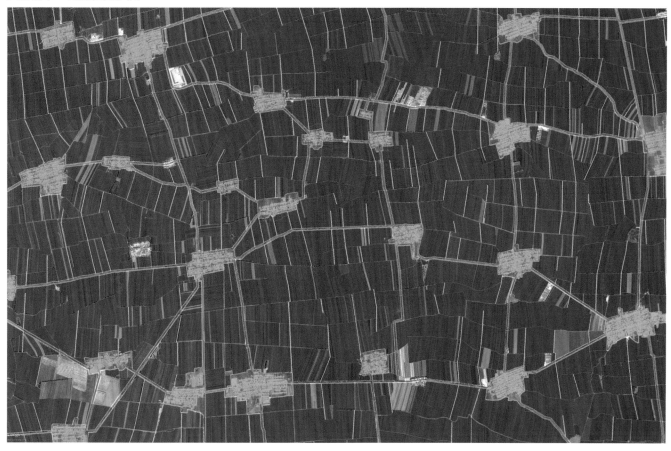

不均匀 / 分散的
UNEVEN OR DISPERSED

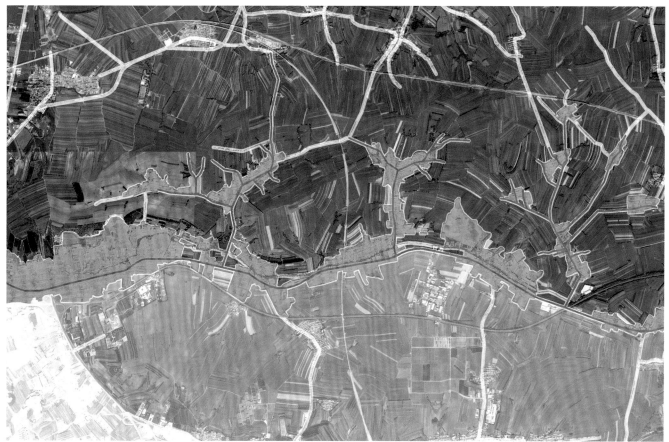

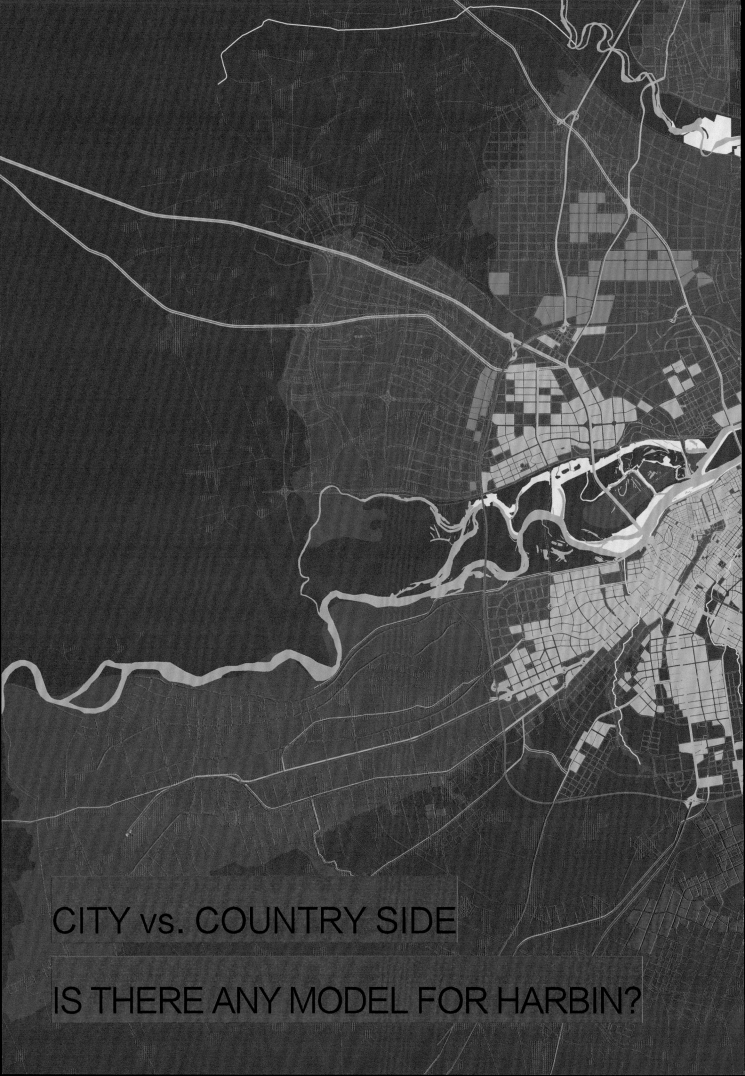

CITY vs. COUNTRY SIDE

IS THERE ANY MODEL FOR HARBIN?

城市 vs. 乡村

哈尔滨发展模式的先例

伦敦
LONDON

伦敦定义了一条绿带
LONDON IS DEFINING A GREEN BELT

巴塞罗那
BARCELONA

巴塞罗那创造了一条山海之间的绿色天然联系
BARCELONA IS CREATING A GREEN NATURAL RELATION BETWEEN SEA AND MOUNTAIN

北京
BEIJING

北京遵循环路的连续发展模式正在尝试定义它的极限
BEIJING CONTINUOUS GROWTH FOLLOWING RING-
ROADS AND IS TRYING TO DEFINE ITS LIMITS

哈尔滨
HARBIN

哈尔滨可以改变其摊大饼式的发展模式。利用目前的基建设施和地理条件给出未来的发展模式，
植根于大都会和农业产业的需求。也许南北向的发展模型可以昭示未来发展的步骤

HARBIN CAN CHANGE ITS ONION GROWING MODEL BY TAKING ADVANTAGE OF THE
CURRENT INFRASTRUCTURES AND GEOGRAPHY TO GIVE ITS FUTURE MODEL
ROOTED WITH THE NEEDS OF BOTH METROPOLIS AND AGROINDUSTRY
PERHAPS NORTH-SOUTH MODEL SEEMS TO SHOW THE FUTURE STEP

绿色区域在城市扩张的系统中作为自然的缓冲带
GREEN AS NATURAL BUFFER IN THE GROWTH SYSTEM

巴塞罗那
BARCELONA

哈尔滨有着欧式的城市肌理。这是一个非常强烈的特点，应当被充分利用

HARBIN HAS A EUROPEAN URBAN TISSUE
A VERY POWERFUL IDENTITY TO MAKE GOOD USE OF IT

哈尔滨
HARBIN

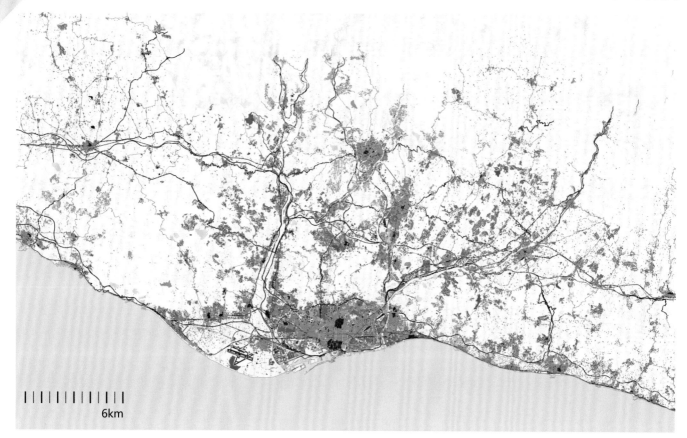

6km

3km

哈尔滨
HARBIN

6km

地理条件与演变

3km

巴塞罗那
BARCELONA

哈尔滨
HARBIN

紧凑城市的品质
QUALITIES OF THE COMPACT CITY

62
63

开敞空间分布图
OPEN SPACE PLAN

哈尔滨
HARBIN

中央大街及其旁支
CENTRAL AVENUE AND BRANCH STREET

道外中华巴洛克
CHINA BAROQUE DISTRICT IN DAOWAI

64
65

南岗花园街地区
GARDEN STREET AREA IN NANGANG

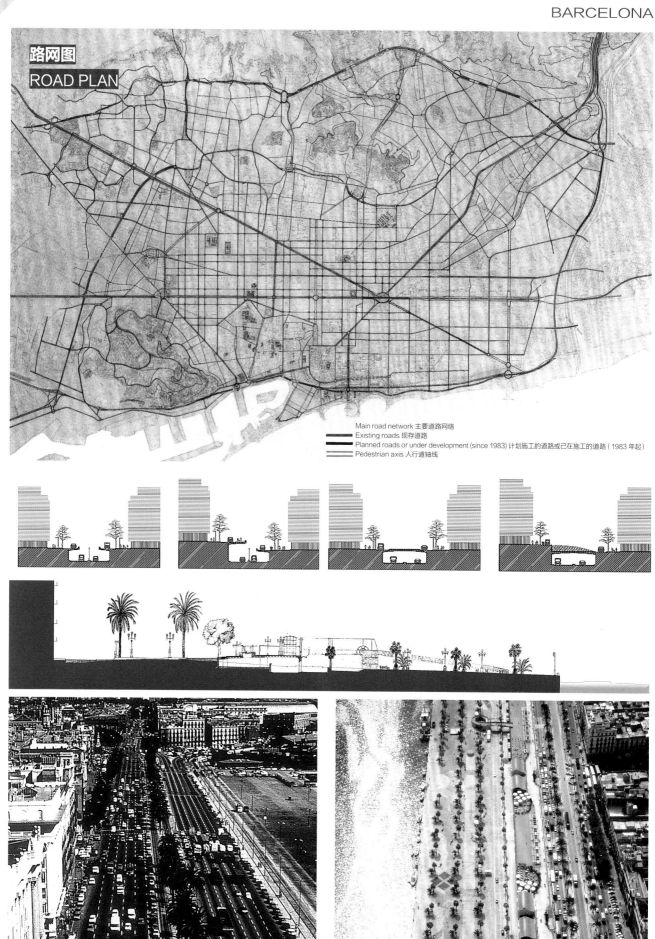

巴塞罗那
BARCELONA

路网图
ROAD PLAN

Main road network 主要道路网络
Existing roads 现存道路
Planned roads or under development (since 1983) 计划施工的道路或已在施工的道路（1983 年起）
Pedestrian axis 人行道轴线

哈尔滨
HARBIN

道路网结构，道路与高速公路

认识道路、高速公路在哈尔滨城市结构中的地位，有利于我们研究向阳镇与城区的关系。直到 20 世纪 80 年代（前面已经阐述过）道路还一直是哈尔滨的城市骨架。

近十年公路开始跨江，这是围绕城市周边的高速环形公路的起点。其他处于城市内部同心圈层的道路，由于地块不连续性或河流的介入而从未闭合。这条 21 世纪建设的环形高速公路，与其他基于城市连续性的大道、现有地块内部道路相比，在城市建设方面是完全独立的。它更符合规划整体布局的要求，尽管与城市本身的功能性没有关系。

哈尔滨高速道路网体系由围合的高速环路与放射状高速公路组成。由于它与城区联系较通达，城市用地可以顺应地形建在高速公路走廊中。然而，它忽略了人口较少的聚落，而且促进了城市扩张。哈尔滨东部的高速公路走廊穿过了向阳镇镇域的部分地区。

The road structure. Roads and highways

The understanding of the foundational structure of paths and from their primary network, helps us understand the structure of the superimposed scale that presents the network of roads and highways. The roads have been Harbin's structural support until the 1980s (explained in the previous chapter) and did not cross the river until this decade. This marks starting point of the construction of the large ring road around the perimeter, concentric to other more interior ones that never got to enclose themselves because of the discontinuity of the tissues or because of the interposition of the river. This ring of highways, practically constructed entirely in this century, as opposed to other interior ones that were based on urban continuities, avenues and existing tissues, is totally autonomous in respect to the city. This ring gets closer to the layout of our project area, although there would be no relationship to functional city with itself.

The peripheral ring roads bypass the radial highways that access the city. This is of greater interest because of how it relates to the territory, and can be placed in the same road corridors that follow these topographic forms. However, it avoids smaller populated settlements and reinforces the urban extension. The eastern corridor crosses the sector of Xiangyang.

当今都市结构已经得到充分的发展
METROPOLITAN ARMATURE IS WELL DEVELOPED TODAY

城市结构对比分析
URBAN STRUCTURE

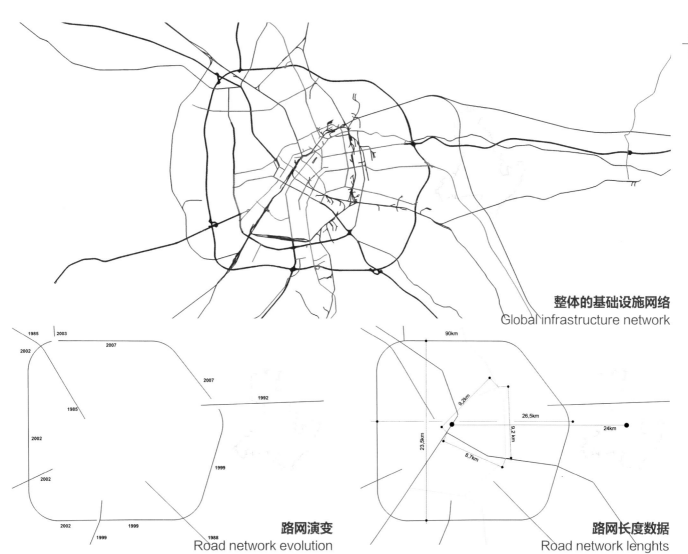

整体的基础设施网络
Global infrastructure network

路网演变
Road network evolution

路网长度数据
Road network lenghts

新的城市中心区
NEW DOWNTOWN AREAS

哈尔滨
HARBIN

地铁

目前，哈尔滨地铁的布局由三条线形成基本结构，并将很快全面运行：

· 线路 L1 在五年前开始通车运行，从南部到东北部穿过城市最古老的部分，没有跨河。

· 线路 L3 环绕市中心，刚刚开通了第一期，并开始运行。

· 线路 L2（正在建设中），从东北向东南穿过河流，我们建议它可以延伸至气象台站，这样就可以与已经规划在向阳镇的高铁接驳。

其余的地铁线路，目前还没有具体的建设日期，它们沿着城市的扩张轴及其临近的区域，呈现出自城市中心向外辐射的布局。

The metro

The metro layout is composed of three lines that form the basic structure that will be in full operation very soon:

· Line L1 became functioning five years ago and travels from the south to the northeast through the oldest part of the city, without crossing the river.

· Line L3 encircles the city center, which recently had its first section open and began operation.

· Line L2 (under construction), crosses the river from the northeast to the southeast. We propose that L2 be extended to reach the Jinshand Rd. Station, which would provide a transfer to the highspeed train and is already inside of the project area.

The rest of the lines, currently without secured dates for their constructions, present a radial scheme coming from the center, following the city's expansion axes and its immediate area.

哈尔滨的基础设施的空间可以被更好地组织起来
HARBIN FACES THE SPACES IN-BETWEEN THE INFRASTRUCTURES WHICH CAN BE BETTER ORGANIZED

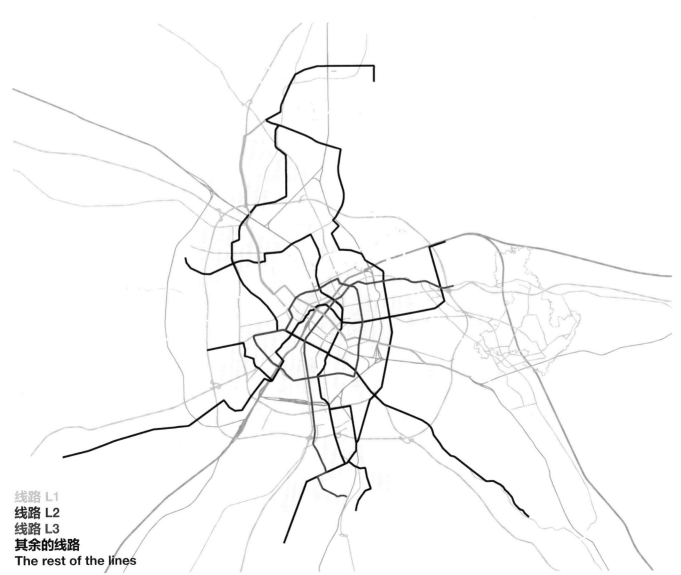

线路 L1
线路 L2
线路 L3
其余的线路
The rest of the lines

乡村空间的形状
THE SHAPE OF THE RURAL SPACE

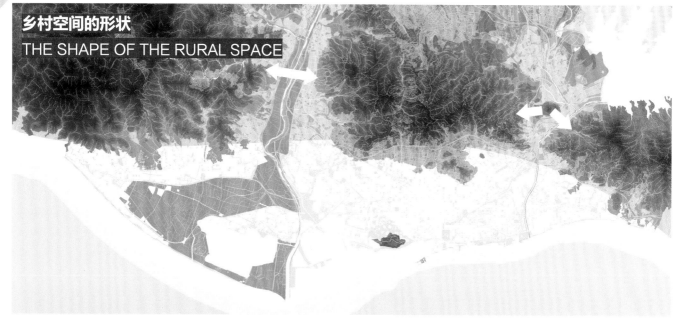

海岸沿线不同的水滨景观
DIFFERENT WATERFRONT LANDSCAPES ALONG THE SHORELINE

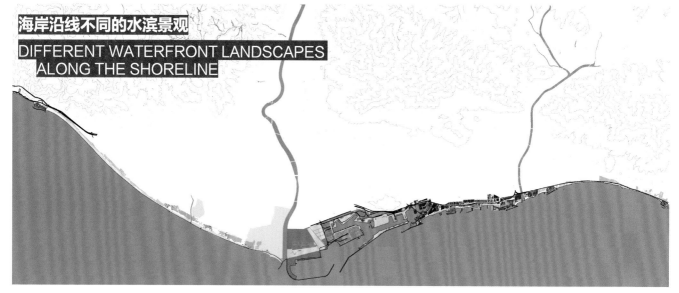

乡村空间的形状
THE SHAPE OF THE RURAL SPACE

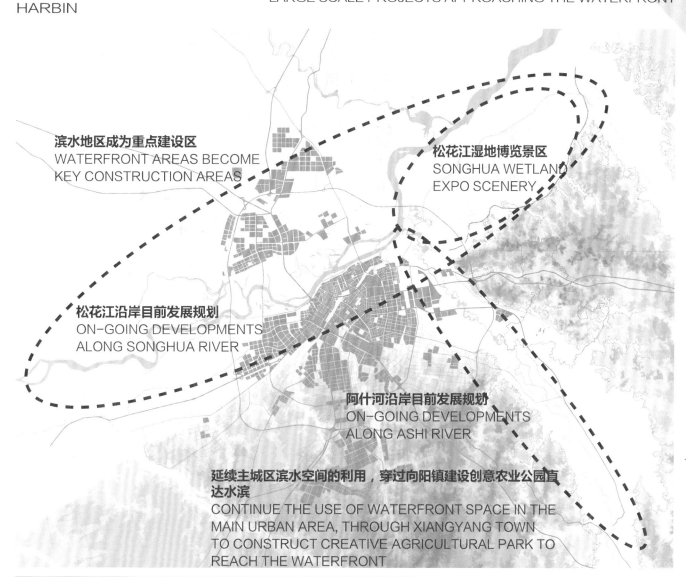

滨水地区成为重点建设区
WATERFRONT AREAS BECOME
KEY CONSTRUCTION AREAS

松花江湿地博览景区
SONGHUA WETLAND
EXPO SCENERY

松花江沿岸目前发展规划
ON-GOING DEVELOPMENTS
ALONG SONGHUA RIVER

阿什河沿岸目前发展规划
ON-GOING DEVELOPMENTS
ALONG ASHI RIVER

延续主城区滨水空间的利用，穿过向阳镇建设创意农业公园直达水滨
CONTINUE THE USE OF WATERFRONT SPACE IN THE
MAIN URBAN AREA, THROUGH XIANGYANG TOWN
TO CONSTRUCT CREATIVE AGRICULTURAL PARK TO
REACH THE WATERFRONT

宏观规划已经在城市区域显现。向阳可作为农业公园的起点
LARGE SCALE INITIATIVES ARE ALREADY HAPPING IN THE METROPOLIS
XIANGYANG IS THE STARTING POINT TO DEVELOP THE AGRO-PARK

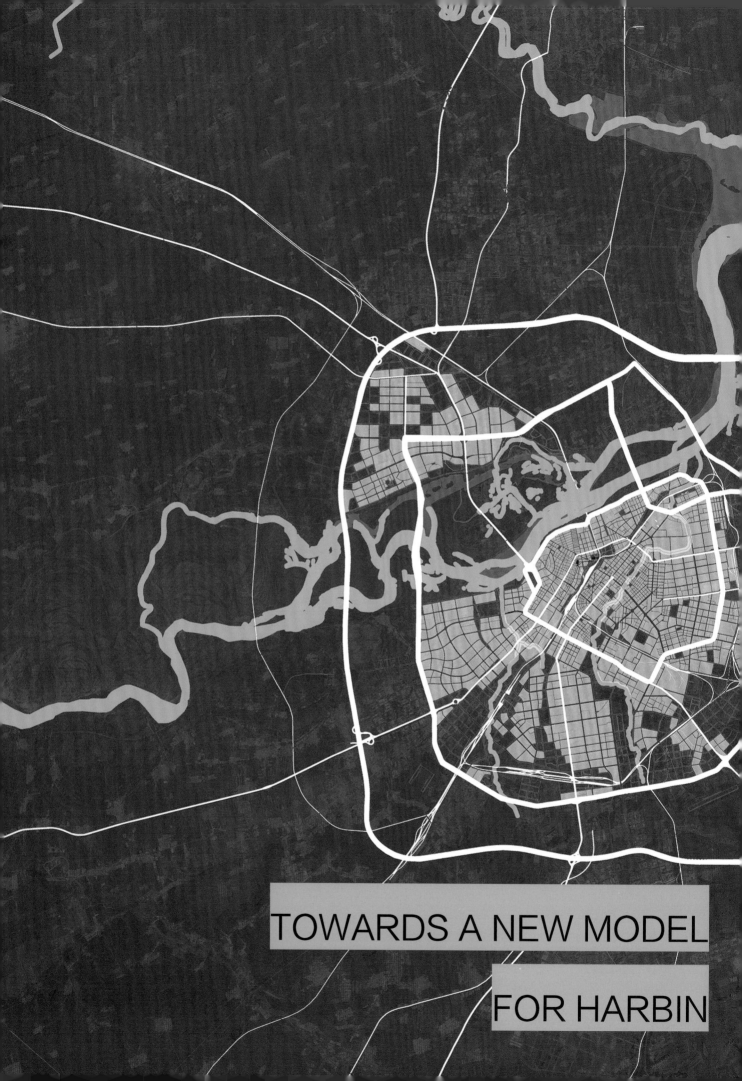

TOWARDS A NEW MODEL

FOR HARBIN

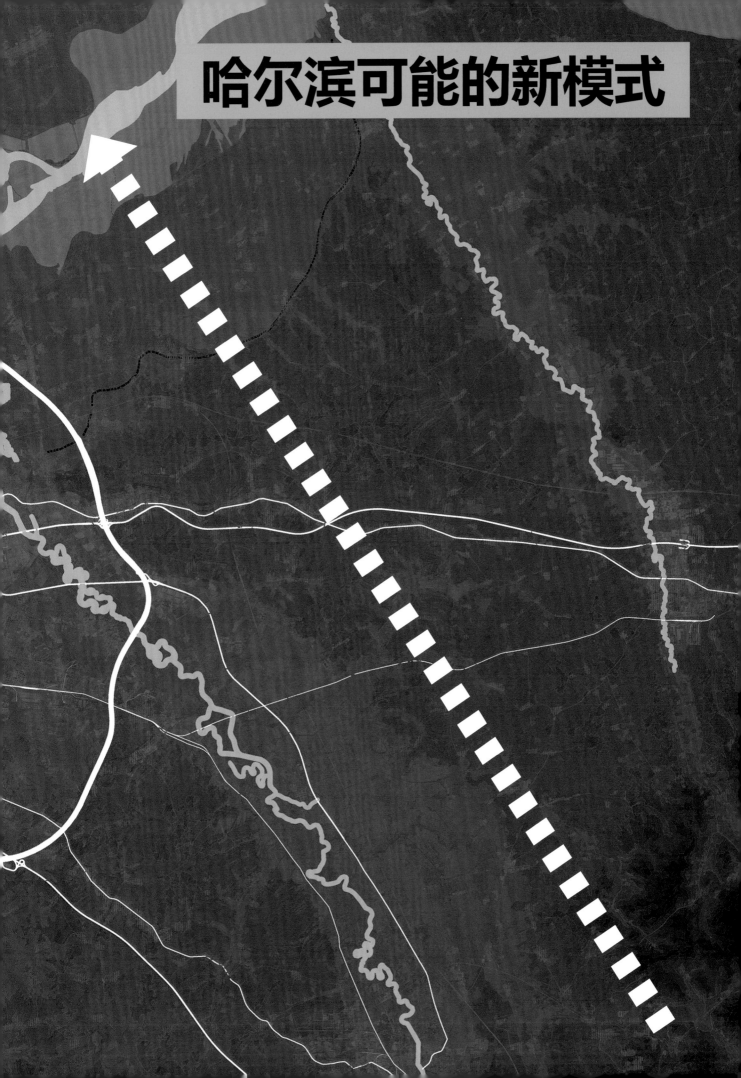

哈尔滨可能的新模式

城市向东部区域的扩张忽略了这一区域的"自然"元素，我们的结论来自城市元素并未与既有的自然空间形态协调以规避城市空间的断裂，抑或是原有土地形态并未被认定为值得考虑的条件。但事实上，这些元素定义了哈尔滨国土空间的特征：

· 划分生物多样性以及植被带的小支流或蜿蜒的溪流

· 形成连续景观的蜿蜒连接的高地与凹谷

· 各种地形元素之间的关系，这种关系创造了谷地与高地交替出现的"手指"，同时线性的高地能更多地嵌入该地域的传统村落。

向阳镇像其他市郊地区一样，得以保留原有的土地形态，保留了相互联系的均质节点和一些谷地的树枝状系统的双重模式，这一切都源于东部大工业区的介入阻碍了所有哈尔滨向东扩张的计划。

认识到并利用这些自然元素的独特性可以作为全新城乡发展模型的基础，将城市连续扩展区域与自然空间区别开来，在具体的城市空间架构内带来革新。将最功能性的城市关系轴线（车、服务、交通、可达性）和最具代表性的城市印象分离开来：环绕流动水体的公园，开敞空间主轴系统，以及在高度开发的城市片区内穿梭的公共交通和可持续交通的通廊。

There exists a tendency to eliminate or hide natural elements of the territory. This is due to the elements not agreeing with the preconceived geometries, to avoid discontinuities or simply to consider their existence as an anecdote that does not deserve consideration. However, these elements characterize the territory:

· Small tributaries or winding streams that travel through the depressions delimiting vegetation belts and biological diversity

· Elevations that form linked meandering grooves creating a continuous view

· The relation between various orographic elements that create "fingers" of alternating valleys with elevations forming a structure in line with what tends to insert the more traditional villages of the territory.

The interposition of the great industrial district to the east has impeded any planned eastern expansion of Harbin towards the east, due to Xiangyang being an opportune territory that already conserves a system of occupation like the rest of the outer territories and maintaining the dual model of interconnected nodes and the dendritic system of some valleys.

Recognizing and utilizing these elements of singularity can be the base of the alternative model that distinguishes the urban expansion of continuity from the natural spaces. These can reform themselves in the specific structure of the urban areas. This can permit the separation of the most functional axes of relation (cars, services, transport, accessibility) in respect to those most representative of the urban image: parks surrounding flowing bodies of water, axes of structure of open spaces and relation between highly constructed and developed pieces with a corridor for public transport and modes of sustainable mobility.

向阳可能的未来

POTENTIAL FUTURE FOR XIANGYANG

向阳镇处于中心城市周边位置，两者之间隔着阿什河，在向阳镇展开建设活动是有问题的（考虑城市肌理的关系方面），或者存在城市发展过于庞大的问题。这里主要是指活动空间、休闲空间、服务空间……它的用途与周边的建成区无关，是基于开放空间的可用性和主要道路的良好可达性而开发的。

将这种不那么有吸引力或太多条件限制的功能放置到城市周边区域的大都市发展设想，给该区域提供了得到良好规划的机会，让建设活动有控制地展开。必须制定一个能最好地整合这些活动的模式，同时考虑到周围区域的特点。

如果前面提到的模式在现状基础设施上实施，那么在研究区域可以实施双重系统的结合，一个系统是将同心圆式发展轴与复杂的更集聚的活动相结合，而另一个则重视传统系统中的土地占有模式，依靠自然、村落和景观价值提高附加值。

The peripheral position of Xiangyang in respect to the central city creates the two problems of the two cities: Xiangyang is already where activities that are considered too problematic (in terms of integration into urban fabric) or too physically large for the developed city become situated. We refer to event spaces, recreational areas, service spaces... Therefore, they are for uses that often have little to do with their developed surroundings and that were developed solely through the availability of open spaces and a good accessibility from the main roads.

However, this metropolitan tendency to place the functions that are less attractive or too complicated on the periphery of the city can be the opportunity to generate well organized areas where these activities can develop in a controlled manner. They have had to formulate a model that best integrates these activities bearing in mind the singularities of the surrounding territory.

If the previously mentioned models correspond to implementations of projects that add to the preexisting infrastructures, the city can implement a dual system in the study area, which would combine a concentric axis of development with complex and more intense activities, with another that values the traditional systems of territorial occupation, in which the natural, rural and landscape values support an added value.

处于一个活力大都市圈边缘的向阳

XIANGYANG AS PERIPHERY OF A DYNAMIC METROPOLIS

提高城市新陈代谢效率的优先权
PRIORITY FOR AN INCREASINGLY EFFICIENT URBAN METABOLISM IN EXISTING CITIES

新型城市规划文化的演变是对一系列元素的响应。通过这些元素，我们可以理解其创新价值以及作为今后城市规划参考的可能性。

21世纪以来，城市规划行业对于环境问题有了崭新的认识。城市规划者在追寻更加可持续的城市设计的过程中，重新发现了以混合功能和紧凑的城市作为核心的传统城市发展模式。同时，人们的关注点放到了低碳聚居点的变化上。以重新紧凑化城市作为城市层面对于可持续化发展的理论手段，还有对于城市规划及其与气候变化的关系之间全新的认识，导致了环境议程的变化，现在它包括公共交通、医疗、贫穷、阶层隔离、公共空间以及内城景观等议题。

我们所熟悉的当代城市建立在一系列实体或者虚拟的网络之上，用以维持其运转以及未来的发展。城市代谢理论让我们可以将城市理解为一个复杂的系统，它同时调试、管理、建构多种资源的积聚和流动，包括能源、水、资源、人、空间和信息。

The evolution of the new urban planning culture corresponds to a series of elements which can help us understand their innovative value and capacity to become references for urban planning practice today and in the medium term.

In the 21st century there is a new understanding of environmental concerns in urban planning. Urban planners have rediscovered the historical model of mixed use and the compact city as a paradigm for more sustainable urban design. Meanwhile, attention has moved towards the transformation of low carbon settlements. Recompacting the city to theorize sustainability at urban level along with the new awareness of the importance of urban planning and its relationship with climate change have resulted in a change to the environmental agenda, which now includes issues like public transport, health, poverty, exclusion, public space and the inner city landscape.

The contemporary city as we know is based on systems of physical or virtual networks that ensure its operation and future transformations. Urban metabolism helps us consider the city as a complex system that calibrates, manages and configures various stocks and flows of resources, such as energy, water, capital, people, space and information.

A great deal of effort is being invested in cities all over the world to develop green infrastructures and restructure the economy towards a future with low CO_2 emissions:

芝加哥低碳化规划
Chicago's Decarbonization Plan

城市代谢与人体的类比
City metabolism as a body

加拉夫（巴塞罗那）垃圾填埋场作为公园
Garraf's (Barcelona) garbage dump as a park

全球各地的城市都在对绿色系统以及低碳排放的新经济进行大量的投入。

比如推广高效的公共交通系统，同时引入丰富的小零售单位以吸引步行人群前往或使用公共交通。

对现有建筑进行现代化改造以及再利用，使用紧凑的城市形态、低影响的公共交通，提供低碳能源，拓宽并延伸公共景观。

绿带和大型公园提高了城市对抗极端气候的适应能力，提高了城市生态上、文化上以及经济上的活力。同时减少不断扩张城市的需求。城市去工业化提供了新的将工业用地转换为城市公园的机会。另外，推广公共交通以及劝阻私家汽车的使用，还可以减少城市中心的停车空间，增加城市公园的可能。城市绿廊能够有效组织景观结构，减少城市当中的热岛效应，改善空气质量，将城市的交通习惯从开车转变为公共交通、步行以及骑行。

– Promoting efficient public transport system and the existence of a rich system of small retailers which invites pedestrians to walk or use public transport.

– With the adaptive reuse and modernization of existing buildings, the compactness of the urban form, low-impact public transport, the supply of decarbonized energy, the occupation boundary and the expansion of public landscape.

– The green belt and large parks provide capacity for resilience achieving ecological, cultural and economic viability. Reducing the demands of constantly expanding cities. Deindustrialization of cities offers new opportunities for transforming industrial land into urban parks. In addition, promoting public transport and dissuading drivers from using private cars allows cities to reduce parking space in the city centre and to turn these areas into urban parks. Green corridors are very useful for structuring the landscape reduce the heat island effect in the city centre, improve air quality and change urban mobility habits from driving to public transport, pedestrians and cyclists.

改进城市与环境的关系
IMPROVING THE RELATIONSHIP OF THE CITY WITH THE ENVIRONMENT

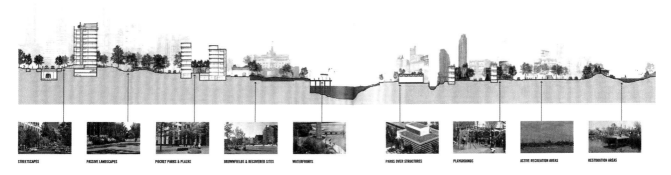

STREETSCAPES   PASSIVE LANDSCAPES   POCKET PARKS & PLAZAS   BROWNFIELDS & RECOVERED SITES   WATERFRONTS   PARKS OVER STRUCTURES   PLAYGROUNDS   ACTIVE RECREATION AREAS   RESTORATION AREAS

纽约开放空间的剖面
New York's open spaces along cross section

New York's ...

Cheonggyecheon ... recovery

大尺度：从松花江到山地
LARGE SCALE: FROM THE SONGHUA RIVER TO THE MOUNTAIN

向阳镇地处哈尔滨东部城市板块与乡村板块交汇处，是城市与乡村过渡带。如何从全局高度提出空间策略方案，是本规划研究的主要目的。

主要目标：以城市设计作为调和自然与工业、居住之间关系的策略。

探索一种特定的自然"智能"（水、地形、植被）来构建一种更广泛的聚合体将是明智的做法，不同的城镇活动和休闲活动都可以在农业自然环境中发生，这将在日益增长的大都市背景下创造衍生式经济效应。

Main goal: Urbanistic design as strategy to reconcile nature with industrial dwelling habits.

It will be wise to explore a specific "intelligence" of nature (water, topography, vegetation) to build a wider composition where agricultural environment can become a common ground where different urban and leisure activities can create economic spin off in a growing metropolis environment.

以城市设计作为调和自然与工业、居住之间关系的策略
URBANISTIC DESIGN AS STRATEGY TO RECONCILE NATURE WITH INDUSTRIAL DWELLING HABITS

哈尔滨东部的潜力

正如对基础设施的历史研究所显现的那样，哈尔滨的扩张是在不同历史层之间的连续性模式之后发生的，构成了一个从非常接近松花江的原点开始扩张的"洋葱式"系统。

现在哈尔滨都市的扩张，正朝着许多不同的方向发展：向北跨松花江的方向变得非常有活力，向西、向南和向东的辐射非常突出。

我们极有可能定义一个不同的"市郊战略"，挖掘哈尔滨东部区域的农业潜力，在阿什河走廊附近以向阳镇区域为主，创建长卷农业公园，也可称为UCAP。

重要的是要考虑哈尔滨周边地区的优良村庄和农业开发网络。这一传统系统具有重要的优势，需要加以保护和增强，以确保中期的良好经济前景。出于这样的原因，大型农业园区的战略可能是一个明智的规划决策，它意味着大都市层面的新城市化结构，意味着整体规模上的新形式形态，与此同时，这一战略可能意味着，类似于过去几十年其他大城市的模式，该地区的农业得以衍生与延伸。

Potentials of the East of Harbin.

As shown in the historical research on infrastructures, Harbin expansion had been happening following a continuity patterns in between the different historical layers, composing a sort of "onion" system from the original center quite close to the Songhua River.

The growth of Harbin Metropolis today is moving in many different directions: The North crossing the river is becoming quite dynamic and the "radius" towards: West, South and East are quite prominent.

It seems quite possible to define a different "Metropolitan strategy" that acknowledging the agricultural potential of the East –considering Xiangyang at large– defining an Agricultural Park as will be described later as UCAP, after the Ashi Corridor.

It is important to consider the excellent network of villages and agricultural exploitation on the surrounding territories near Harbin. This traditional system has an important strength that needs to be protected and enhanced to ensure a good economic future in the mid-run. For such a reason the strategy of a large Agricultural Park could be an advisable planning decision that implies new urbanistic structure at the level of Metropolis, –meaning a new formal morphology for the overall scale; at the same time this strategy can mean a spinoff for the Agriculture in the region following the patterns of other large Cities in the last decades.

农业聚居点对于泛都市区的农业产业以及经济来说都是非常重要的
AGRICULTURAL SETTLEMENTS ARE VERY POWERFUL FOR THE AGROINDUSTRY AND THE ECONOMY OF THE METROPOLIS

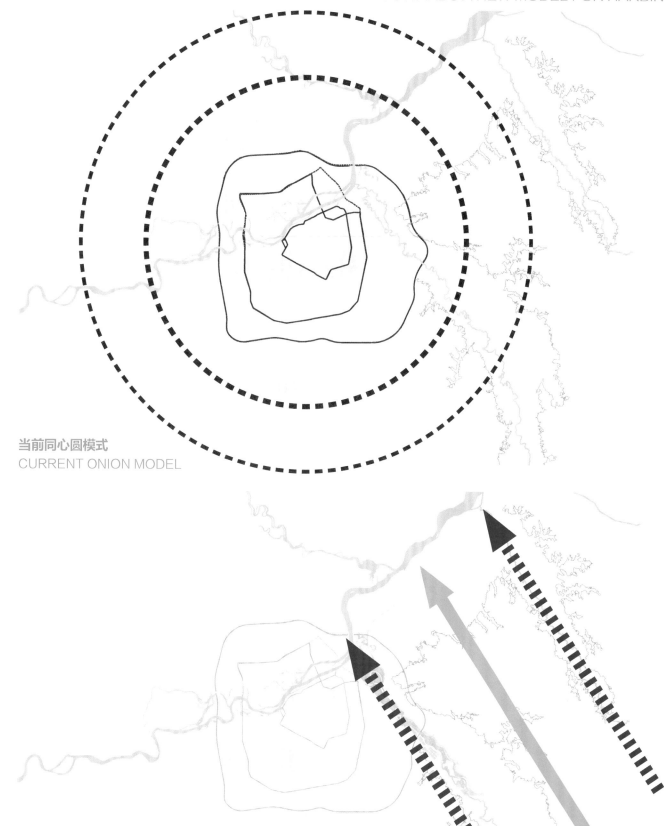

哈尔滨可能的新模式
TOWARDS A NEW MODEL FOR HARBIN

当前同心圆模式
CURRENT ONION MODEL

未来模式
FUTURE MODEL

未来将山体系统与水体连接的模式

THE FUTURE MODEL JOINS THE MOUNTAINOUS SYSTEM WITH THE FLUVIAL

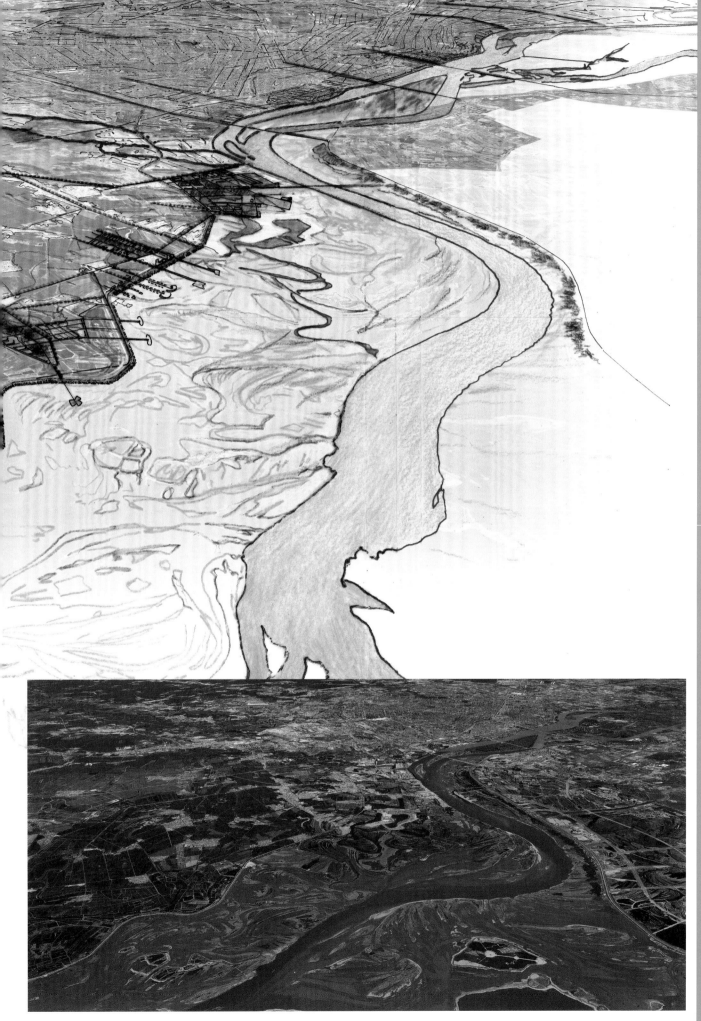

INDUSTRIAL DEVELOPMENT

STRATEGY

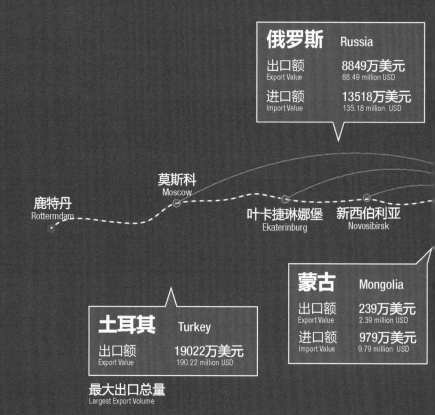

俄罗斯 Russia

出口额 8849万美元
Export Value 88.49 million USD

进口额 13518万美元
Import Value 135.18 million USD

莫斯科
Moscow

鹿特丹
Rotterdam

叶卡捷琳娜堡
Ekaterinburg

新西伯利亚
Novosibirsk

蒙古 Mongolia

出口额 239万美元
Export Value 2.39 million USD

进口额 979万美元
Import Value 9.79 million USD

土耳其 Turkey

出口额 19022万美元
Export Value 190.22 million USD

最大出口总量
Largest Export Volume

最大进口总量
第二大贸易总量
Largest Import Volume
Second Largest Total Trade Volume

巴西 Brazil

进口额 65882万美元
Import Value 658.82 million USD

1. AGRO-INDUSTRY

2. LEISURE AND TOURISM

3. SERVICES FOR RESIDENTS

产业发展战略

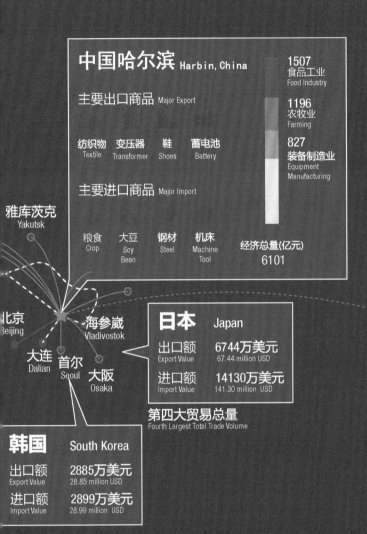

中国哈尔滨 Harbin, China

主要出口商品 Major Export

纺织物	变压器	鞋	蓄电池
Textile	Transformer	Shoes	Battery

主要进口商品 Major Import

粮食	大豆	钢材	机床
Crop	Soy Bean	Steel	Machine Tool

1507 食品工业 Food Industry

1196 农牧业 Farming

827 装备制造业 Equipment Manufacturing

经济总量(亿元) 6101

雅库茨克 Yakutsk

北京 Beijing

海参崴 Vladivostok

大连 Dalian

首尔 Seoul

大阪 Osaka

日本 Japan
出口额 **6744万美元** Export Value 67.44 million USD
进口额 **14130万美元** Import Value 141.30 million USD

第四大贸易总量 Fourth Largest Total Trade Volume

韩国 South Korea
出口额 **2885万美元** Export Value 28.85 million USD
进口额 **2899万美元** Import Value 28.99 million USD

进出口贸易总量最大 Largest Total Trade Volume

美国 The United States
出口额 **10144万美元** Export Value 101.44 million USD
进口额 **58141万美元** Import Value 581.41 million USD

洛杉矶 Los Angeles

货物运输量 (万 t) Goods Transported (x 10000 tons)

铁路 Rail — 785
陆路 Road — 7492
水路 Water — 431
航空 Aire — 4

图标 Legend
—— 现有货运航线 Air Cargo Route (Existing)
---- 规划货运航线 Air Cargo Route (Planning)
- - - 第一亚欧大陆桥 Eurasian Land Bridge

1. 农业产业

2. 休闲旅游

3. 本地服务

目前，向阳镇乃至中国城市近郊农村，整体所面临的社会和经济问题，大多源自农业生产效率低下，收入过低，无法与大城市丰富的就业机会和高水平的工资竞争，造成人口流失。因此，如何提高农业生产效率，提高农业收入，改善农村收入结构，从而吸引农村人口回流，是产业规划当中的重点课题。

在这一章里，我们将深入细致地研究向阳镇目前的经济基础，以判断小镇发展的优势和劣势，给出向阳镇的战略定位，推断规划给经济带来的影响。

向阳镇内功能的演变及哈尔滨的影响

本地已经有了一些新的功能项目。我们可以看出城市已经认识到向阳镇距离城市较近的优势，正在这里寻求新的用地。而向阳镇的农业氛围正在努力抵御城市扩展的侵蚀，然而创新力不足。

我们对诸项目进行认真研究，避免功能和空间上的重复，以丰富向阳镇内的活动和体验，形成项目之间的协同作用。总体而言，向阳镇内当前的开发项目以休闲旅游为主，本地公用设施建设几乎不存在；农业产业，特别是高价值农业产业，对周边带动辐射作用不强。新的结构性规划应当帮助农业产业品质的改良和创新，同时具有足够的开放度以迎接新型经济活动的产生。因此，我们需要一种新的规划方式。

产业协同力理论

经验表明，一个地区经济的发展不能仅仅依靠内部的资源自生驱动，外界的投入至关重要。然而如何有效地利用外部资源驱动内部产业系统，形成可持续性的经济发展模式，则需要合理地安排农业产业、休闲旅游业以及本地建设之间的关系，利用彼此之间的协同作用，避免重复性功能类型之间的相互竞争。这样才能用更少的外部资源达到更多的经济发展效果。例如，休闲旅游需要外部的投资和当地的劳动力吸引城市居民和外国游客的消费，同时可以作为当地农业产业的窗口吸引商贸客户。农业产业的增收可以为本地村屯带来更高的收入，改善当地的学校，培养更多的高技能劳动力，从而提高当地农业产业和休闲旅游的效率。

产业规划中需要考虑不同功能类型对于用地、投资和劳动力的不同需求。农业产业通常需要大面积的场地和高昂的资金设备投入，同时需要高技能的多样化的人才团队和本地劳工的配合，需要大量的销售运输等辅助设施，有着很高的准入门槛。休闲旅游功能对于场地和资金投入较为灵活，对于劳动力的要求也不高。本地建设主要靠当地公共和私营部门的累计收入以及当地劳动力来改善当地居住和生活条件。

Main social and economic problems of Xiangyang Town as well as rural villages on the urban fringe in China result from low agricultural productivity and low income. The inability for rural villages to compete with major metropolis in terms of job opportunities and income lead to loss of labor in rural area. Therefore, the key economic planning in rural China is to improve agricultural productivity, increase farming income, and improve income structure, so that rural population will return to the countryside.

This chapter will study the economic fundamentals of the town from various scales and perspective. It will serve as the basis for us to evaluate development strength and weakness, determine strategic position of the town, and deduce the effect of our proposal on the economy.

Evolution of functions in Xiangyang today. Metropolitan influence from Harbin

We have found out that there are many new programs already happened in Xiangyang town. We can see that agricultural atmosphere in Xiangyang is resisting without many innovations but there is a process of development of new metropolitan activities searching for new land and taking advantage of the proximity to the large city of Harbin.

We briefly studied these programs to avoid functional and spatial repetitions, to enrich the activities and experiences which Xiangyang will offer and form a synergy between programs. Generally, new programs in Xiangyang are dominated by leisure tourism; local settlements have barely any public amenities; and agro-industry, especially high-value agro-industry, hasn't been able to project influences to the surroundings. It seems clear that new Structural Project should try to improve and innovate the quality of the agro-industry at the same time that can allow new economic activities to happen. For this a new type of planning seems to be needed.

Synergy

Experience has taught us that the economic development of a region cannot solely rely on its own resources and the organic economic mechanism. Nevertheless, in order to utilize external resources to drive the local economy and form a sustainable model of development, we need to carefully arrange the placement of agro-industry, leisure and tourism, and local settlement to form economic synergy and avoid un-fair competitions from repetitive functions. Thus, we will be able to leverage less external investment to achieve greater economic success. For example, leisure tourism development requires external investment and local labor to attract metropolitan residents and foreigners to consume. Meanwhile, it can be used as a window for local agro-industry businesses to attract clients. As agro-industry witness greater revenue, local settlements will have more economic resources to improve local school and nurture more skilled labors, who in turn will improve the efficiency of local agro-industry and leisure tourism.

Program development also needs to consider the different requirements for different functions, including space, capital investment, and labor. Agro-industry often needs large open spaces, heavy capital investment, highly skilled talent team and also not well-trained labor, and supplementary facilities for trade and transport. The barrier of entry is high for agro-industry. On the other hand, leisure tourism is much more flexible regarding space, capital investment, and labor. Local settlement improvements rely on the accumulative earnings from both the public and private sector and local labor to improve living conditions.

大面积土地
Large Area

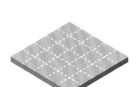

多元人才
Diversified Talents

农业产业
Agro-Industry

农产品
Produce

重资产投入
Major Capital
Investment

农业产业需要更多高技术工人以及投资，从而带动当地的经济以及环境的品质

AGRO-INDUSTRY DEMANDS MORE SKILLED WORKERS AND INVESTMENTS, THAT WILL RISE ECONOMY AND QUALITY OF THE AREA

常规劳动力
Common Labor

开放土地
Open Land

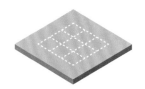

休闲旅游
Leisure/ Tourism

娱乐
Entertainment

资产投入
Capital Investment

新的社区服务将会吸引新的居民

NEW SERVICES FOR THE RESIDENTS WILL ATTRACT NEW RESIDENTS

常规劳动力
Common Labor

建筑空余
Infill

本地村落
Local Settlement

积蓄
Accumulated
Earnings

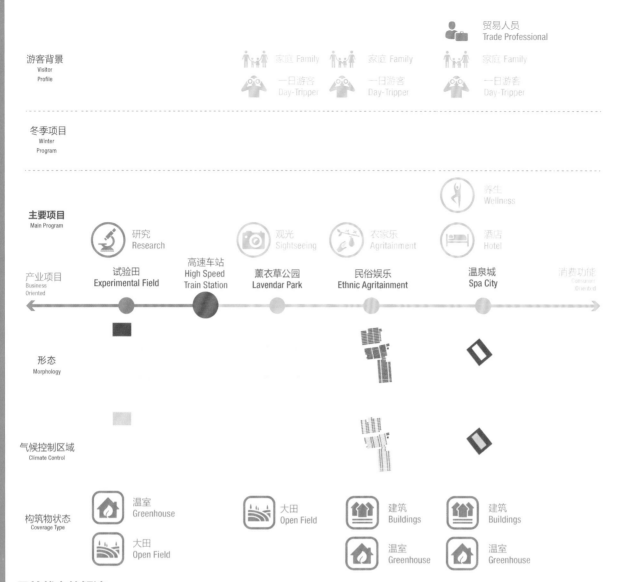

游客背景
Visitor Profile

贸易人员
Trade Professional

家庭 Family 家庭 Family 家庭 Family
一日游客 Day-Tripper 一日游客 Day-Tripper 一日游客 Day-Tripper

冬季项目
Winter Program

主要项目
Main Program

养生
Wellness

研究
Research

观光
Sightseeing

农家乐
Agritainment

酒店
Hotel

产业项目
Business Oriented

试验田
Experimental Field

高速车站
High Speed Train Station

薰衣草公园
Lavendar Park

民俗娱乐
Ethnic Agritainment

温泉城
Spa City

消费功能
Consumer Oriented

形态
Morphology

气候控制区域
Climate Control

构筑物状态
Coverage Type

温室 Greenhouse
大田 Open Field
大田 Open Field
建筑 Buildings
温室 Greenhouse
建筑 Buildings
温室 Greenhouse

目前状态的解读
INTERPRETATION OF EXISTING CONDITION

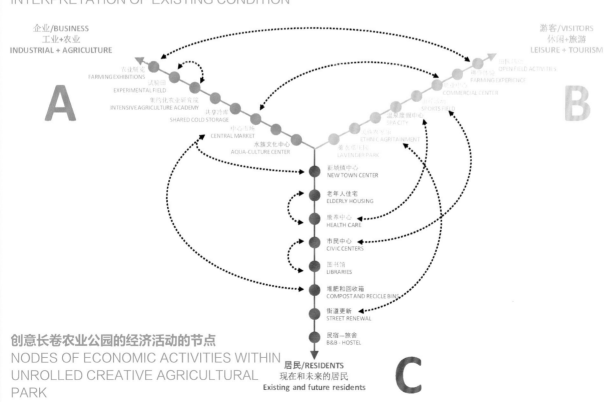

企业/BUSINESS
工业+农业
INDUSTRIAL + AGRICULTURE

游客/VISITORS
休闲+旅游
LEISURE + TOURISM

A

B

C

农业展览
FARMING EXHIBITIONS
试验田
EXPERIMENTAL FIELD
集约化农业研究院
INTENSIVE AGRICULTURE ACADEMY
共享冷库
SHARED COLD STORAGE
中心市场
CENTRAL MARKET
水族文化中心
AQUA-CULTURE CENTER

开放体验
OPEN FIELD ACTIVITIES
采摘体验
FARMING EXPERIENCE
商业中心
COMMERCIAL CENTER
运动场
SPORTS FIELD
温泉度假中心
SPA CITY
民俗农家乐
ETHNIC AGRITAINMENT
薰衣草公园
LAVENDER PARK

新城镇中心
NEW TOWN CENTER
老年人住宅
ELDERLY HOUSING
康养中心
HEALTH CARE
市民中心
CIVIC CENTERS
图书馆
LIBRARIES
堆肥和回收箱
COMPOST AND RECICLE BINS
街道更新
STREET RENEWAL
民宿—旅舍
B&B - HOSTEL

居民/RESIDENTS
现在和未来的居民
Existing and future residents

创意长卷农业公园的经济活动的节点
NODES OF ECONOMIC ACTIVITIES WITHIN UNROLLED CREATIVE AGRICULTURAL PARK

根据可能吸引的人流，经济活动可分为三种类型——

面向商业：主要是工业和农业活动与设施。

面向游客：与休闲和旅游活动有关。

面向居民：在现有定居点内或附近提供设施。

虽然可能需要拓展新的土地，但建议将它们放置在现有的村落附近并诱导其转变。此外，开放场地活动等公共活动场所可以在定居点附近或其他用地附近，以促进功能混合。

Economic activities are divided into 3 types according to the kind of people that they will attract or help:

–Business oriented: Mostly industrial and agricultural activities and facilities.

–Visitors oriented: Related to leisure and tourism activities.

–Residents oriented: To provide facilities inside or nearby the existing settlements.

Although some might demand new land, it is recommended to place them near the existing settlements to induce its transformation. Also, edifications related to activities such as open filed activities can be nearby the settlements or other uses in order to boost the mixed-use.

运作机制
OPERATIONAL MECHANISM

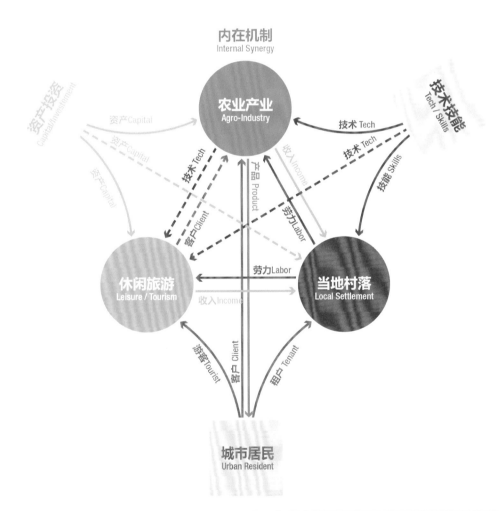

外部资源
External Resources

内在机制
Internal Synergy

两种经济支点已经存在，必须加强第三个经济支点：针对居民的服务，以激活三大主要经济支点所形成的合力，改善大都市区的环境

TWO ECONOMIC BRANCHES ARE ALREADY IN PLACE. WE MUST ENHANCE THE THIRD: SERVICES FOR THE RESIDENTS, TO PRODUCE SYNERGIES BETWEEN BRANCHES AND IMPROVE THE METROPOLIS

案例研究的分类
CATEGORIES FOR THE CASE STUDIES

区域和城市政策：

· 农业旅游：小规模的如莱切或佩内德斯或大规模的类似加利福尼亚州纳帕谷的案例。

· 健康、运动和休闲。

· 郊区食品供给：戴尔略弗雷特三角洲，米兰。

· 城市食品供给（分散）（慢食）。

城市和市区改造：

· 农业填补城市空白（底特律）。

· 大型公园附带有活力的农业。

· 绿色走廊；激活废弃的场所（亚特兰大环线）。

· 景观基础设施（雨水）。

建筑物和周边特殊区域：

· 集约农业（荷兰的温室）。

· 水产养殖（印度尼西亚，鱼米养殖场）。

· 结合自然风光和休闲活动的"主题公园"（华特迪士尼等）。

· 垂直农业，多层次的发展。

· 独立型：休闲旅游热点周围并未出现休闲旅游产业集聚，也未依赖其他功能吸引游客。

· 核心型：休闲旅游热点吸引了大量人流，并激发了周围小型休闲产业的集聚。

· 挂靠型：项目借助周边大型吸引中心所带来的人流成为休闲旅游热点。

· 自发型：项目周边具有类似大小及功能的自发产业聚落，在共同吸引人流的同时彼此竞争。

· 规划型：规划利用集聚效应，在一定空间里组织差异化的产业功能项目，发挥多功能协同作用扩大吸引力。

Territorial and Urban Policies：

–Agritourism: Small scale in Lecce or Penedes or large-scale system like Napa Valley, California.

–Health, Sport and Leisure.

–Nutrition and Metropolis: Delta Llobregat, Milan, ...

–Cities and Nutrition (scattered). (slow food)

City and Urban retrofitting：

–Agriculture filling the urban voids. (Detroit)

–Large park with active agriculture.

–Green corridors; activating abandoned sites. (Atlanta Beltline)

–Landscape infrastructures. (storm water)

Buildings and Special Precincts：

–Intensive agriculture. (Glasshouses in the Netherlands)

–Aquaculture. (Indonesia, combining rice and fish farms)

–"Theme park" combining nature and leisure. (Walt Disney, etc)

–Vertical agriculture. Multilayer developments.

–Stand-alone: The project hasn't activated any functional cluster, nor does it depend on other developments to attract visitors.

–Anchor: The project acts as an anchor to attract a large volume of visitors and activates smaller developments nearby.

–Spill over: The project capitalizes on the traffic generated by nearby anchor and become a popular site.

–Cluster: The project is surrounded by similar projects in size and function. The organic cluster works together to attract visitor while also competes against each other.

–Planned: The planned development leverages the cluster effect of industry and orchestrates a group of diverse programs to maximize the attraction.

分类: Categories:	农业旅游 Agritourism	郊区食品供给（系统） Nutrition and Metropolis (Systems)	城市食品供给(分散) Cities and Nutrition (Scattered)	大型农业公园 Large Parks with Agriculture	绿道联系 Green Connectivity	雨水 Stormwater
产业集群类型: **Aggregation or Compositional type.**						
独立型 Stand-alone			阿尔博温室 Arbor House 鹰街屋顶农场 Eagle Street Rooftop Farm 新式农场 The New Farm 弗赖汉姆区 Freiham District	皇后郡 Queens County	环线 The Belt Line	西部公园 Westside Park
核心型 Anchor	维拉弗兰卡和佩内德斯 Vilafranca and Penedes 萨伦提娜乡村 Salentina 普里奥拉 Priorat 托斯卡纳小镇 Campagna Toscana	城市增长边界 Urban Growth Boundary 寿光蔬菜基地 Shouguang Vegetable Base 官渡蔬菜基地 Guandu Vegetable Base 大爱生态农场及养老中心 Da Ai Eco Farm & Elderly Care Center		总督岛 Governors Island 兰德尔岛 Randall's Island 弗莱士河公园 Freshkills Park 济南植物园 Jinan Botanical Garden	滨水绿道 Blue Greenway 434号公路 Route 434 翡翠项链公园 Emerald Necklace Park	弗雷泽河要地 Point Fraser
挂靠型 Spill over	Napa Valley 纳帕谷			总督岛 Governors Island 兰德尔岛 Randall's Island		
自发型 Cluster	卡希纳农舍 Cascina Farmhouses					
规划型 Planned		戴尔略弗雷特三角洲 Delta del Llobregat 帕尔科阿古利可拉南部 Parco Agricolo Sud	五个自治市 Five Boroughs	弗莱士河公园 Freshkills Park	乔路易斯绿道 Joe Louis Greenway	

市镇发展状态:
Urban Condition:
市区 / Urban
郊区 / Metropolitan
农村 / Isolated

以下是来自世界不同地区的 50 个成功案例研究的摘要。

这是根据规划规模、构成类型和城市条件进行案例研究分类编组的。

此外，气候已被考虑在内，尽管大多数案例可能在不同的气候地区进行。

这些提供了一些思路来处理管理、规划、更新、公共政策、复兴和发展战略等问题。

The following is a summary of 50 successful case studies from different territories around the world.

Those are organized in a "matrix of case studies" according to theirs: Scale-program; compositional type; and urban condition.

Also, the climates have been taken into account although most of the initiatives could take place in different climate areas.

Those, brings several clues to arrange issues such as: management, programming, renewal, public policies, rejuvenation and development strategies, among many others.

集约化农业 Intensive Agriculture	水产养殖 Aquaculture	主题公园 Theme Park	健康与运动 Health and Sport	农家乐 Agritainment	农业填补空白 Agriculture and filling the voids	垂直农场 Vertical Agriculture
		皮尔斯伯里滨河公园 Pillsbury Riverfront Park 埃姆舍公园 Emscher Park 郁金香主题公园 Tulip Theme Park 稻梦空间 Rice Fantasy 百万葵园 Million Sunflower Garden		龙裕山庄 Longyu Villa		纽约垂直农业 NYC Vertical Agriculture 淡水基金会 Sweet Water Foundation
北京草莓博览会/农业嘉年华 Beijing Agri Carnival 十方缘葡萄生态园 Shifangyuan Grape Ecological Park	稻—鱼养殖 Rice-Fish Farming	马灿 Marzahn 现代都市农业博览会 Modern Urban Agriculture Expo Park		木兰胜天农庄 Mulan Shengtian	底特律未来城市计划 Detroit Future City Plan	
		大顶子山度假村 Dadingzishan Spa				
玻璃温室作物体系 Glasshouse Food System 顺利草莓采摘园 Shunli Strawberry Picking		世界园艺博览会 World Garden Expo		妙杳生态农庄 Miaoxing Ecological Farm	底特律未来城市计划 Detroit Future City Plan	
	赞比亚水产养殖 Zambia Aquaculture		新西兰 New Zealand		特殊时期农业改革 el Periodo Especial	

美国加利福尼亚州纳帕谷
Napa Valley, California, USA
轴线作为区域的"脊柱"
Axis as a "Backbone" for the territory

加拿大安大略省伦敦舞茸菌菇培育所
Shogun Maitake, London, Ontario, Canada
恶劣天气中萌发的经济类型
A new economy in difficult weather

美国佐治亚州亚特兰大市环线;预计2030年完成
The Belt Line, Atlanta, GA, USA; 2030
绿道提升连通性
Green axis to improve connectivity

美国纽约州皇后郡农场博物馆:1975年
Queens County Farm Museum, NY, USA; 1975
传授可持续的耕作技术
Teaching sustainable cultivation techniques

西班牙巴塞罗那戴尔略弗雷特三角洲

Delta del Llobregat: Barcelona, Spain

保护大都市区的自然和农业遗产

Preservation of natural and agricultural heritage in the metropolitan region

德国鲁尔地区埃姆舍公园; 1989—1999年

Emscher Park, Ruhr, Germany;1989–1999

分期开发的景观公园

Landscape Park Developed in Phases

荷兰玻璃温室作物体系

Glasshouse Food System, Netherlands

气候控制，"玻璃温室"的"规模"问题

Weather control. "sea of glass"

意大利米兰帕尔科阿古利可拉南部；1990年

Parco Agricolo Sud, Milan, Italy;1990

保护农业遗产和大都市区食品体系

Preserving agricultural heritage and metropolitan food system

传统农业
Conventional Farming

有机农业
Organic Farming

220hm² 种植农场
Arable Farm

 饲喂技术
Animal Feeding
X 140

放牧技术
Grazing
X 60

120hm² 牛奶农场
Dairy Farm

 X 40
养马场 Horse Farm

 X 1000
蛋鸡场 Poultry Farm

 X 280
养猪场 Pig Farm

实践基地
Practice Academy

 AERES　艾尔斯农业公司
Aeres Landbouwonderneming

 AERES
AGREE　雇佣平台
Job Platform

 Agro *transfer*　商务中心
Business Center

  学生公司
Student Company

PAARDENPLAATS
AERES HOGESCHOOL DRONTEN

营收服务
Income-Generating Serivce

实习经营训练学生实践经验
同时获得免费劳动力
Intership management
training students with
practical experience while
gaining free labor

与农业大学课程紧密相连，
提高课程效果
Closely linked with
agricultural university
courses to improve course
effects

引入农业商务中心
吸引规模企业同时孵化农业创业
Introduce an agricultural
business center to attract large-
scale enterprises and incubate
agricultural entrepreneurship

教学经营并重的
农业学校

Agriculture Academy for
Education and Enterpreneurship

艾尔斯农业学校实践基地 | 德隆特，荷兰
Aeres Praktijkcentrum | Dronten, the Netherlands

建造年份：2005
Year Built
总面积： 340hm²
Total Area

学生数：10
Student
年盈利：€ 78000
Annual Profit

雇员数： 5
Employee
运营主体：公私合营
Management Public-Privaste

　　艾尔斯是荷兰最为重要的农业教育集团。2015 年，艾尔斯拥有近 1100 名员工，1 万名学生，资金流水近 1 亿欧元。艾尔斯实践基地为艾尔斯教育集团下属的各部分校区提供农业食品行业的短期职业培训。

　　实践基地主要由一个农场和学生公司组成。农场当中分为耕地和三个不同的牛奶农场。其中一个针对动物饲喂，一个针对自动化畜牧，还有一个针对放牧。在此之外，这里还有一个马场、一个养猪场、一个养鸡场。农场的耕地由艾尔斯下属农业企业运营。园区内设立了农业商务中心，针对规模企业和创业公司，同时还为社会性的展示和训练提供场地。

Aeres is one of the most important agriculture education entity in the Netherlands. In 2015, it had approximately 1100 employees, 10000 students. The turn-over is close to €100 million. Aeres Praktijkcentrum Dronten offers short term practical training for professionals in the agro & food sector.

It has a farm and student companies. The farm of consists of an arable farm and three different dairy farms: one focusing on animal feed and research, one focusing on automation,and one focusing on grazing. In addition, there is a horse farm, a pig farm and a poultry farm. The arable farm of Aeres Farms consists of a conventional biological company with its own mechanization and storage of products. Furthermore, the center is used for demonstrations and specific training courses for business.

数据来源 Data Source
1. 艾尔斯集团 2017 年年报。
　 Aeres Group. Annual Report 2017.
2. www.aeresfarms.nl

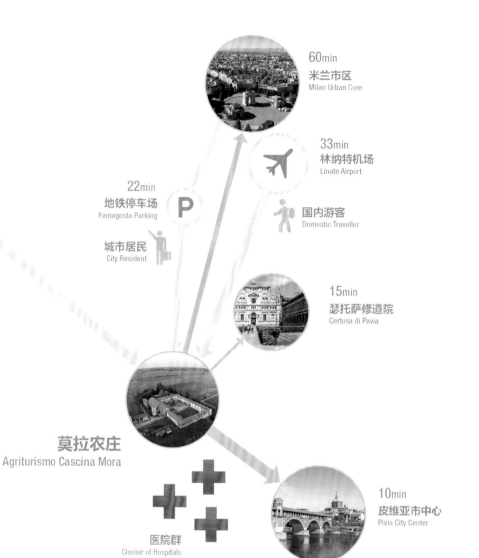

60min
米兰市区
Milan Urban Core

33min
林纳特机场
Linate Airport

22min
地铁停车场
Famagosta Parking

城市居民
City Resident

国内游客
Domestic Traveller

55min
马尔彭萨国际机场
Malpensa Airport

国际游客
International Traveller

15min
瑟托萨修道院
Certosa di Pavia

莫拉农庄
Agriturismo Cascina Mora

10min
皮维亚市中心
Pivia City Center

医院群
Cluster of Hospitals

多样化稻米产品
网上直销最大化价值
Diversified Rice Products
Sold Online To Maximize Value

传统田园风光
可用于长期停留
Traditional Countryside Scenery
Prepared For Extended Stay

当地农民迁入城市后
闲置房屋用于住宿及幼儿园
Local Farmers Moved Away
Emptied Rooms Are Used For
Hospitality And Kindergarten

稻米种植与农庄风光
Rice Procution Meets Agriturismo

莫拉农庄 | 帕维亚，意大利
Agriturismo Cascina Mora | Pavia, Italy

建造年份：2011 Year Built	价格区间：$79~$111 Price Range	运营主体：私营 Management: Private
总面积：82hm² Total Area	房间：9 Rooms	雇员：4 Employee

45%
意大利 Italy

35%
欧盟其他 Other EU

20%
世界其他 Other

Book.com旅客来源
Book.com Guest Source Data

　　莫拉农庄建立在一个 15 世纪农庄的基础之上。20 世纪 70 年代以后，现代的耕作方式减少了对劳动力的需求，农民离开土地搬到了城里，农业活动就此停止，年轻的萨瓦尼夫妇从 2007 年起重新开始了农业生产，并于 2011 年开设了民宿。由于良好的服务和无与伦比的环境，民宿从意大利商业部获得了"意大利优质酒店"的称号。同时，多样化服务与产品帮助农庄取得了更好的经济效益。农庄周围环绕着 82hm² 的农田，种植稻米、玉米和大豆。民宿的客人可以参与到农业生产的各个环节中来。作物的种植仍然依赖 16 世纪设计的灌溉网络。在这片地区仍然可以找到罗马时期的遗迹。而且在这片土地上，农业景观和乡镇环境之间的交流非常明显。

Agriturismo Cascina Mora is buit upon a historical site that can be dated by to the 15th century. Agricultural production was active untile 1970s, when agricultural production became less labor intensive and farmers moved to the city. The young Savini couple restarted the farm business in 2007 and established the agritourismo in 2011. The hotel received great success for its friendly service and exceptional environment, receiving the honor "Ospitalità Italiana di Qualità" from the Chamber of Commerce. Moreover, the diversification of services and product strengthens the farm's financial standing. The farm is surrounded by 82 hectares of rice, corn and soybeans. Our guests can attend all the phases of the agricultural campaign. Crops are still strongly connected to the use of the water network designed especially in the 16th century. Traces of the Roman age have been found in the area. And the exchange between the rural and urban reality is evident in the territory.

数据来源 Data Source
1. http://www.risocascinamora.it/about/
2. https://www.cascinamora.it/

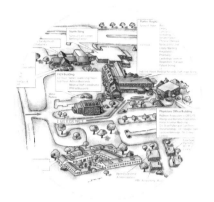

与自然和乡村景观融为一体
有益于身心健康
Integrated With Natural And Rural
Landscape, Which Is Good For Health

毗邻区域性医疗中心
便于求医就诊
Abutting Regional Medical Center
Convenient for Resident
To Seek Medical Help

周边农场直接供应新鲜农产品
保证健康同时提振当地产业
Fresh Produce Supplied By Nearby
Farm To Ensure Healthiness

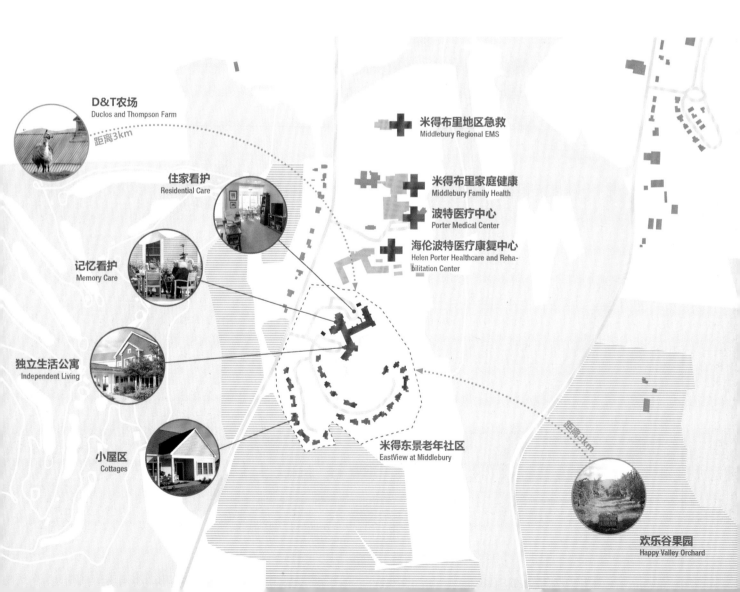

D&T农场
Duclos and Thompson Farm

距离3km

住家看护
Residential Care

记忆看护
Memory Care

独立生活公寓
Independent Living

小屋区
Cottages

米得布里地区急救
Middlebury Regional EMS

米得布里家庭健康
Middlebury Family Health

波特医疗中心
Porter Medical Center

海伦波特医疗康复中心
Helen Porter Healthcare and Rehabilitation Center

米得东景老年社区
EastView at Middlebury

距离3km

欢乐谷果园
Happy Valley Orchard

医疗中心旁的老年社区
Elderly Care by a Medical center

米得东景退休社区 | 米得布里，美国
East View at Middlebury | Middlebury,U.S.A

建造年份：2012
Year Built

总面积： 16hm²
Total Area

资产：$27467082
Asset

收入：$5530039
Income

运营主体： 非营利机构
Management: Nonprofit organization

住户： 105
Resident

　　米得东景是一个本地人自发组织的非营利退休社区，这里的环境非常适合作为退休社区：良好的大学城、美丽的景观以及多元的文化背景。这个社区提供 30 个小屋、31 个独立生活公寓、20 个居住看护公寓以及 18 个记忆看护房间，并提供短期病床看护。截至 2013 年 6 月，超过105 名年逾 62 岁的居民入住米得东景。

　　整个社区与该地区的医疗中心接壤，包括波特医疗中心和海伦波特医疗康复中心。这对于居民寻求及时的医疗救治来说是很方便的。附近的农场直供给社区新鲜的本地农产品，保证食物质量的同时形成了具有持续性的商业模式。

East View at Middlebury is a locally-governed non-profit retirement community in Middlebury. The location is perfect for such a program: the remarkable college-town setting, the beauty and charm of the landscape, and the diverse cultural offerings in the region. The complex provides 30 cottage homes, 31 independent living apartments, 20 residential care apartments, and 18 memory care studios, and respite care. By June 2013, more than 105 residents, all above 62 years old, called East View their home.

The community is abutting the regional medical center including Porter Medical Center, Helen Porter Healthcare and Rehabilitation Center, which is very convenient for the residents to seek medical care in time. Nearby farms also directly supply the community with fresh local produce, ensuring the quality of food and also creating a sustainable business.

数据来源 Data Source
1.http://www.eastviewmiddlebury.
com/

空间规划策略

SPACE PLANNING STRATEGY

1. AGRO-PARK BOUNDARY

2. CIVIC AXIS

3. ACTIVITY NODES

4. RETROFITTING EXISTING VILLAGES

1. 农业公园边界

2. 公众轴线

3. 活动节点

4. 现有村落改造

向阳镇策略方案：城市架构
XIANGYANG AT LARGE STRATEGIC PROPOSAL: URBAN ARMATURE

城市架构作为创意长卷农业公园整体项目的初始步骤的呈现，我们可以对讨论中出现的变化进行相应操作，以作为其实施的测试。

它的简要描述是：

①巩固现有的城市架构。比如对如何扩展一些道路，以提供更好的汽车和卡车等交通服务做一个详细的假设，等等。此外，改善居民的公用设施和电线网络。一些设施——如市场、学校、体育设施、小型办公室等——可以按照前面章节中的拟案进行规划。

②用市民轴线定义绿色基础设施。绿化主轴在真实的地形和土地覆盖范围内限定。绿化和聚集空间的序列可以预期。

③开发少数经济活动战略节点作为整体的样本范例。在向阳镇，应该选择在创意长卷农业公园中更有可能开始吸引新活动的战略节点。活动节点为：向阳镇创新农业研究教学中心；同时也吸引哈尔滨大都市区的居民的本地居民的食品市场；拥有良好服务的温室，它可以在这种气候下试验更高效的农业，同时可以吸引更远方的人流；等等。

虽然每个战略节点都是针对特定活动而组织的，但这些节点不应该是封闭的区域，相反，它们必须整合其他用途，以适用于不同类型的用户（商人、游客和村民）并充分利用地理位置优势。例如：温室节点可以是农业产业系统的一部分，既可用于生产蔬菜，也可用于直销，其中一部分还可用于教学目的，并且其位置使其可以为游客提供观光平台和穿越田野的休闲路径。

④村庄。对现有村庄进行的研究确立了它们不同的类型和类别。从这一点开始，根据城市架构和市民轴线的位置，我们可以轻松选择它们扩展的可能性和服务的优先级。这些扩建和翻新应该使用新的住房类型和用途，主要目的是留下当前的村民并吸引新的居民。

⑤将阿什河绿色走廊和高铁之间的"紧凑型城市"正式化。这个部分包括正在发展的文化城市。沿着哈尔滨郊区的城市形态非常密集，边缘（界限）非常明确；正如我们在巴塞罗那的戴尔略弗雷特三角洲公园中所展示的。一些手指状的绿化渗透可以在创意长卷农业公园和阿什河走廊之间建立联系。

It can be presented as initial step of the UCAP overall project and can be operated with the changes that will appear in the discussion, as a test for its implementation.

A brief description of it could be:

– Consolidating the existing Urban Armature. A detailed hypothesis of how some roads can be extended to provide better service to mechanical flows of cars and trucks, etc. Also, improving the network of utilities and wires for the residents. Some amenities –like markets, schools, sports facilities, small offices, etc– can be planned as the simulations in previous chapter.

– Defining the Green Infrastructure with the Civic Axis. Green spine is defined in the real topography and land coverage. Sequence of green and gathering spaces is envisioned.

– Development of few strategic nodes of Economic activity as samples for the overall. In Xiangyang District should be selected the strategic Nodes that are more capable to start attracting new activity to the UCAP concept. Activity nodes as: research and teaching center to start innovative agriculture in Xiangyang; food market for residents but also attracting people from Harbin metropolis; glass-houses with good services to experience more productive agriculture in this climate, can be attractor for people for far distance; among others.

Although each strategic node is meant for a specific activity, those should not be closed precincts, on the contrary, they must integrate other uses for different kinds of users (Business, Tourists and Villagers) and taking advantage of the geography. For example: a glass-house node can be part of the agro-industry system for both producing vegetables but also to direct-selling, part of it could be used for teaching purposes and its position allows it to include sight-seeing platforms and cross-county path for leisure.

– Villages. Research done on existing villages establish diverse types and categories. From this and according the position of the armature and the location of the Civic Axis we can easily select the extension possibilities and service priorities. Those extensions and refurbishments should look for new housing typologies and uses, mainly keep the current villagers and to attract new citizens.

– Formalizing a "compact city" in between the Ashi green corridor and the HST train. This sector includes the going on development cultural city. Morphology is quite dense following patterns of outskirts of Harbin and edges (limits) are quite well defined;as we can show in Delta Llobregat Park in Barcelona. Few green fingers can make the links between UCAP and the Ashi corridor.

1 农业公园边界
Agro-Park Boundary

2 公众轴线
Civic Axis

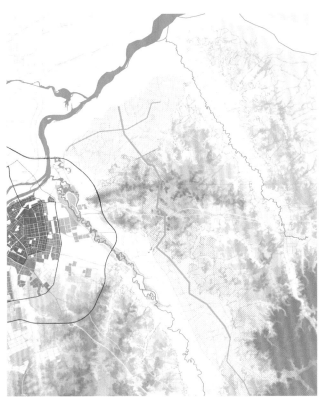

3 活动节点
Activity Nodes

4 现有村落改造
Retrofitting Existing Villages

农业公园确定了管理的边界。公众轴线连接主要片区与单一活动节点，同时村落得到了改善
AGRO-PARK DEFINES BOUNDARY TO BE MANAGED, CIVIC AXIS LINKS MAIN PIECES,
WITH SINGULAR ACTIVITY NODES, WHILE VILLAGES ARE IMPROVED

公众轴线主要截面
CIVIC AXIS' MAIN SECTION

夏季功能设想
HYPOTHESIS
FOR SUMMER

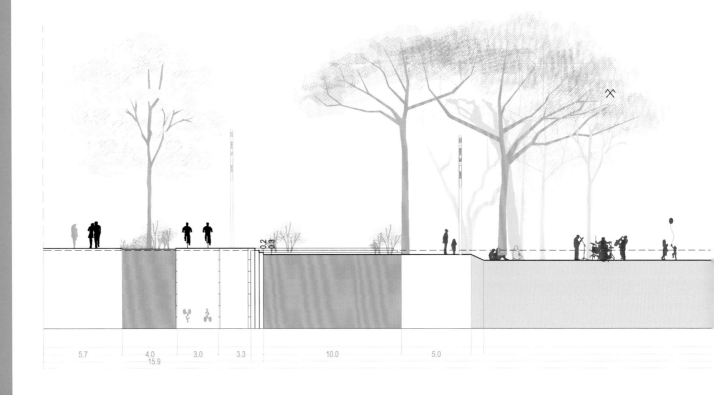

5.7   4.0   3.0   3.3   10.0   5.0
15.9

冬季功能设想
HYPOTHESIS
FOR WINTER

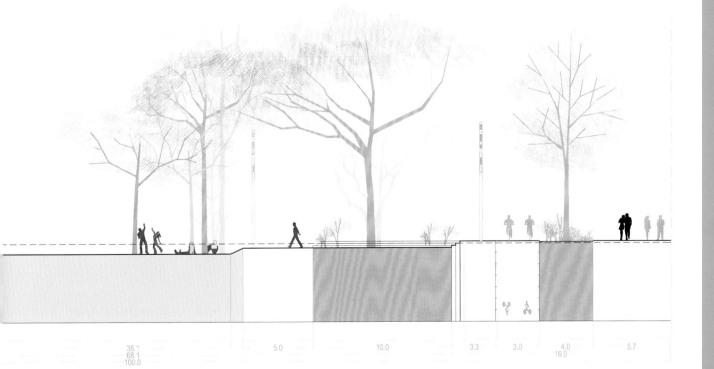

38.1
68.1
100.0

5.0   10.0   3.3   3.0   4.0   5.7
16.0

城市架构：节点的双重接口
URBAN ARMATURE: Nodes' Double Access

总体战略规划依赖于精心设计的"场所"选择，而场所是由"模式"系统设计的。

——在这一重要的总图规划中，可以进行多种方案模拟。这确保了提案的合理性，使得发展策略可以随着时间的推移而发展和转变。

为了理解一系列"模式"如何发展，必须关注"形态学"。它们解释了如何填补方案，因为其他项目的经验可以根据经济回报以及要创建的环境类型用于借鉴。

形态学是正式的和结构性的组织，它实现了不同的建筑类型和不同的项目的开发。它们由建筑物与开放和／或集体空间组成。一般来说，它们允许许多不同类型的开发。

城市形态有时会"呈现"更广阔或更重要的城市构成策略。它们可以保证：均质性；正面与背景；地标战略；高铁等产生的特殊节点。

这里包括一些典型的"特殊城市节点"。它们阐述了"城市公共空间作为在自然环境中创造品质的主要手段"的重要性。

Master plan strategy counts with the well-crafted selection of "places" to be developed, and they are designed by a system of "patterns".

–In this master key scheme, many simulations are possible. This ensures the rational proposal for the development that can evolve and be transformed along the time.

To understand how a series of "patterns" can be developed, "Morphologies" are presented. They explain how proposed patterns can be filled: Experiences from other projects then can be applied in terms of economic return as well as the type of environment to be created.

Morphologies are formal and structural organizations that allow different architectural typologies and different programs to be developed. They are composed by buildings and open and/or collective spaces. In general, they allow many different types of developments.

Urban morphologies are sometimes "submitted" to wider or more important strategies for Urban Composition. This may guarantee: homogeneity; front versus background; landmark strategy;special nodes because of HST, etc.

Also some paradigmatic "special urban nodes" are presented. They recall about the importance of the "urban public space, as main device to create quality in this natural environment."

新的节点功能必须完善现有用途并防止重复
NEW NODES FUNCTIONS HAVE TO COMPLEMENT EXISTING USES AND AVOID REPETITION

每个节点都与发展轴线和道路网相联系
EACH NODE IS LINKED TO THE CIVIC AXIS AND CONNECTED TO A ROAD

很多人会希望在乡间生活，同时还能与城市中心取得良好的联系
MANY PEOPLE WILL WANT TO LIVE IN THE COUNTRYSIDE, WELL
CONNECTED TO THE CENTER OF THE METROPOLIS

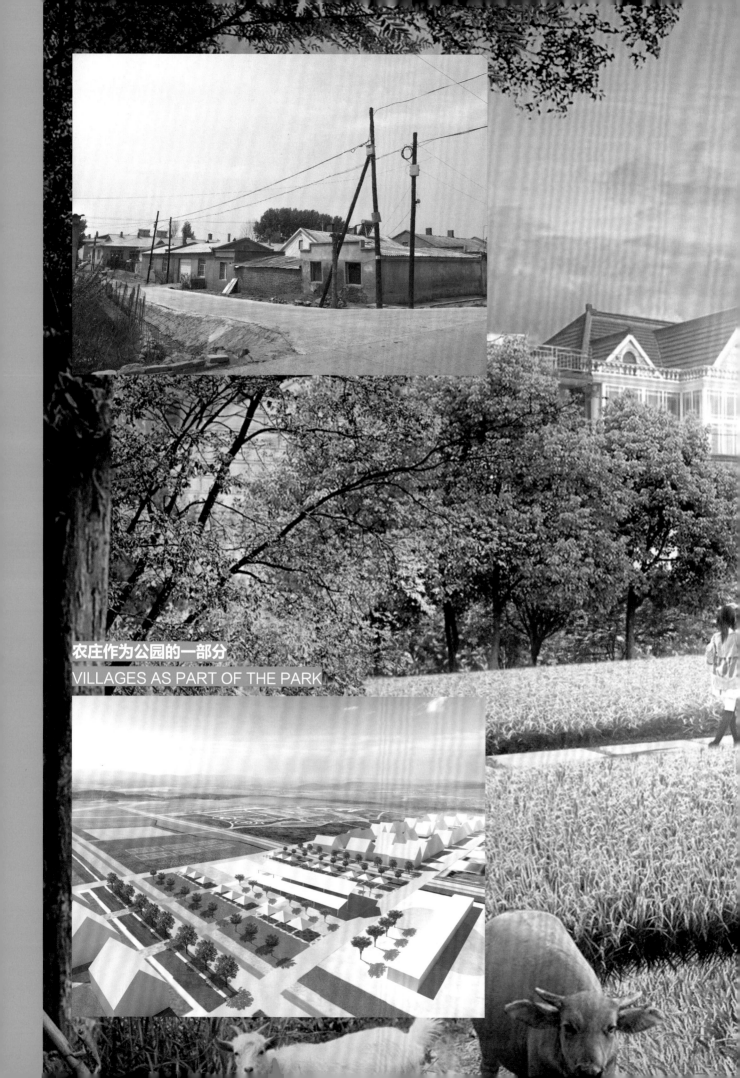

农庄作为公园的一部分
VILLAGES AS PART OF THE PARK

通过改善村庄，居民将会意识到经济的改善
BY IMPROVING VILLAGES, RESIDENTS NOTICE THAT ECONOMY IS MOVING ON

实施推广策略

IMPLEMENT PROMOTION

STRATEGY

阿什河 Ashi River

杨洪业灌渠 Yanghongye Irrigation Channel

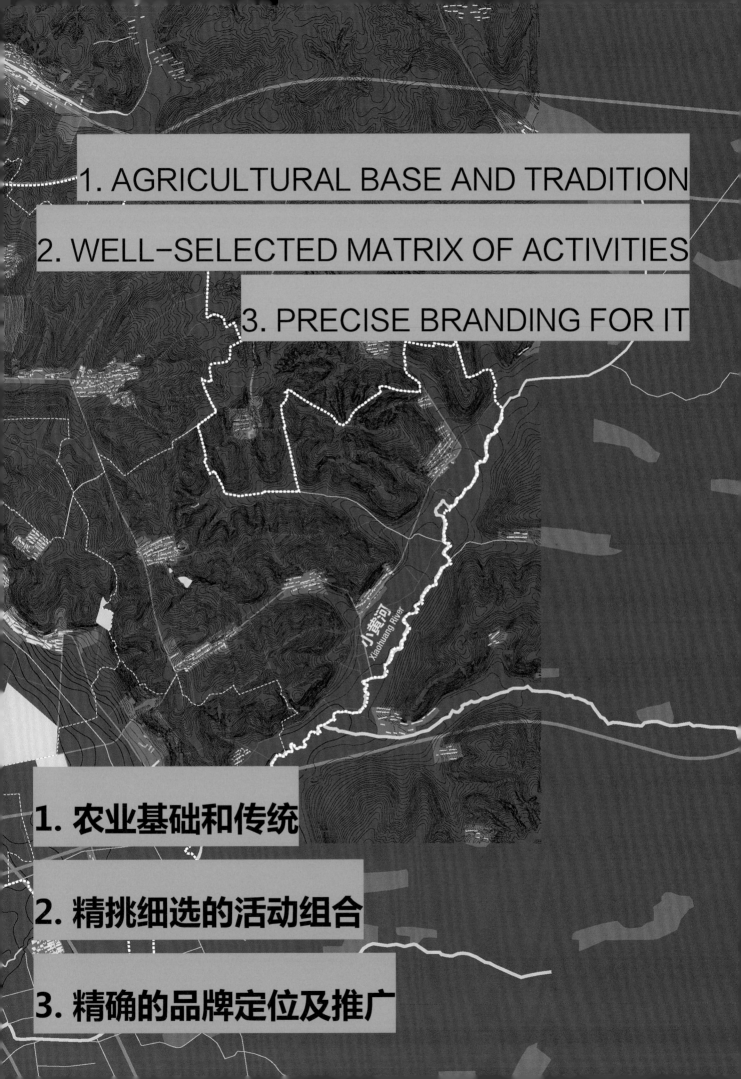

1. AGRICULTURAL BASE AND TRADITION

2. WELL–SELECTED MATRIX OF ACTIVITIES

3. PRECISE BRANDING FOR IT

小黄河
Xiaohuang River

1. 农业基础和传统

2. 精挑细选的活动组合

3. 精确的品牌定位及推广

现有人口 **10550**
Population

住宅用地： **215** hm²
Residential

现有产业用地： **179** hm²
Existing Business

其他： **1411** hm²
Other

华南城 Huanancheng Development

预计人口： **30000**
Projected Population

住宅用地： **100** hm²
Residential

产业用地： **300** hm²
Business

其他： **100** hm²
Other

2018 种子批发公司
Seed Wholesale

2003 向阳工业小区
Xiangyang Industrial Zone

新香坊北
Xinxiangfangbei

天汇小镇 Tianhui Cultural Town

预计人口： **600**
Projected Population

住宅用地： **50** hm²
Residential

产业用地： **240** hm²
Business

其他： **10** hm²
Other

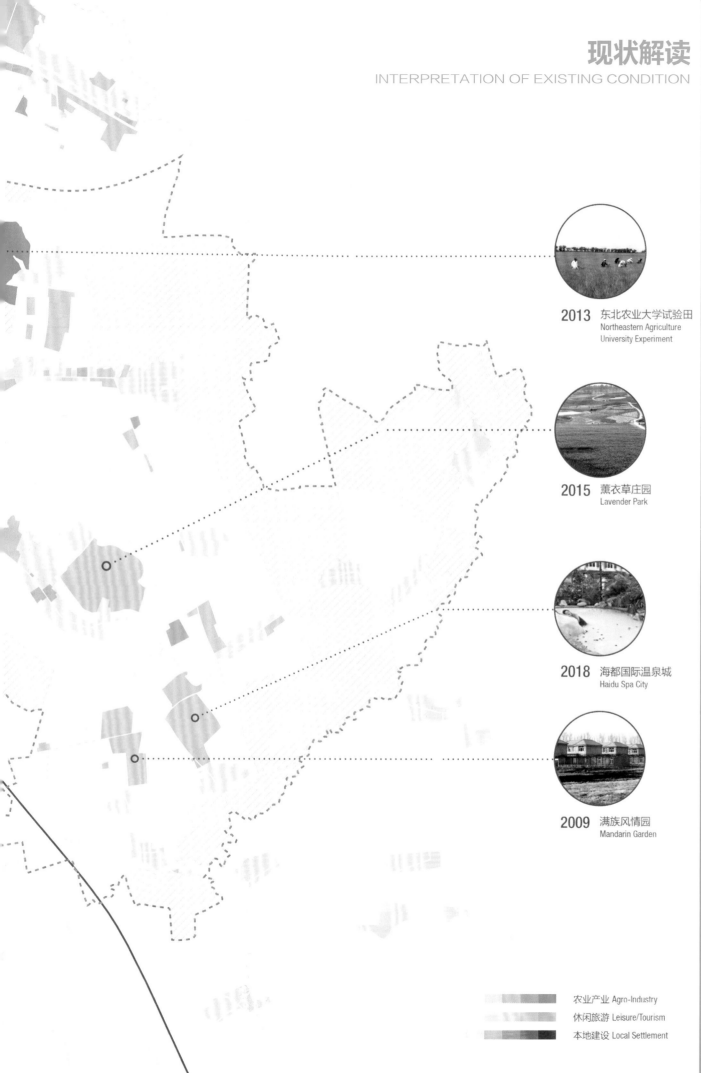

2013 东北农业大学试验田
Northeastern Agriculture
University Experiment

2015 薰衣草庄园
Lavender Park

2018 海都国际温泉城
Haidu Spa City

2009 满族风情园
Mandarin Garden

农业产业 Agro-Industry

休闲旅游 Leisure/Tourism

本地建设 Local Settlement

1988

1998

2008

2016

114
115

概述

向阳特色小镇由五个行政村落和一个社区组成。五个行政村落内下属 23 个自然村屯，在镇域内相对平均地分布，彼此独立。东方红社区前身为国营东方红农场及其配套职工社区设施，与行政村落有着不同的管理政策。由于地形、交通、适用政策、民族结构的不同，向阳特色小镇镇域内的村镇有着多样化的形态，可以提供丰富的生产、生活、生态体验。

数据统计

由于哈尔滨市区在小镇西侧，特色小镇村屯内的建筑密度由西向东逐渐降低，户均耕地资源则由西向东逐渐升高。小镇当中东西向的满福街，也是丘陵和平原地区的分界线，在镇域内人口分布最为密集，包括石槽村和东方红社区。其中，东方红社区的户均人口仅为 2.0，远低于镇域平均值（3.4），更类似于城市家庭人口结构，而不像是村镇的。

村屯现状

向阳镇直到 2013 年都没有统一规划，乡镇建设完全由当地人自发组织。受到宅基地政策、土地承包政策、传统耕作半径的共同影响，村落分布相对密集，而单一居民点规模较小，平均每个自然村屯有约 106 户人口，耕地分割相对碎片化。

自发的村屯建设受当地经济基础的影响很大。总体来说小镇南部的平原地区和西侧靠近长江路和江南中环路的地区，经济基本面富足，村屯建设面貌良好，有明显的自我更新能力。而东部、北部的丘陵地带，村屯环境品质较低，缺乏改造动力，需要外界的介入。

组织结构

特色小镇域内自然村落大小不一，最小的东兴村李太屯 62 人，最大的石槽村杨洪业屯 1278 人。因此，村庄内部的组织结构也不尽相同。根据观察，组织结构主要有平行、垂直、树状和网格四种类型，主要由建筑朝向的要求和地形决定。由于强烈的季风气候影响，中国北方地区整体对于坐北朝南的建筑朝向要求非常严格。同时，多数村落都依照地形建立在丘陵的低洼地带，不轻易占据高地。

村落形态

村屯的形态有助于我们解读村落的建造历史和决策背景。根据观察，我们将村屯内院落的形态分为五大类：单一型、背靠背型、双入口型、大型院落型以及住宅小区型。目前村屯当中，每个院落大小在 600~900m²，远大于黑龙江省新建宅基地标准（350m²）。随着农村建设用地政策的放开，今后较大的院落有可能承载商住混合型功能。

Overview

Within the special town boundary, there are five administrative villages and one community. In the five administrative villages, there are 23 independent organic villages evenly distributed within the town area. On the other hand, Dongfanghong community has inherited the former state-run Dongfanghong farm and its associated facilities, which means the community is governed under different land laws from the villages. Due to the differences in topography, infrastructure, applicable law, and ethnic structure, the village morphology is highly diverse, which will lead to rich experiences.

Statistics

Since Harbin urban center is on the West, building density within the town area decreases from West to East, while the average farmland area per household increases. Manfu Street, the west-east division between the hills and the plain, has concentrated the densest residential belt in the town area, including Shicao village and Dongfanghong community. Moreover, the average population per household in Dongfanghong community is as low as 2.0, which is far below the town's average (3.4) and much closer to an urban family structure than a rural one.

Conditions

Historically, Xiangyang town didn't have a master plan until 2013. Village and town building has been completely self-organized. Because of the influence from housing policy, land contract policy, and working radius for traditional farming, the distribution of villages are relatively dense while each residential settlement is small, averaged at 106 households per village. Consequently, the farmland is fragmented.

Self-organized village building is greatly affected by the local economy. In general, villages in the plain area in the south and the ones close to Changjiang Road and Jiangnan Central Ring Road in the west are relatively well-off. They have a better environment, and they have shown the capability to self-renovate. Meanwhile, villages in the hilly area in the East and North suffer from poor building conditions and the lack of momentum for rejuvenation. Outside intervention is needed in these areas.

Structure

The organic villages vary in sizes. The smallest village only has 62 people while the largest one has 1278. Therefore, the organizational structure within each village is also different. Based on our observation, there are four major types of village structure: parallel, perpendicular, tree-shape, and network. The structure type is determined mostly by building orientation and topography. As northern China is dominated by Monsoon climate, south-facing building orientation is an almost unbreakable rule for rural housing. In terms of topography, most of the villages are built on the lower land within the hilly area.

Morphology

The study on village morphology is helpful for our reading of the villages' building history as well as their background. Based on our observation, we categorize the plots in five types: Suburban, single-access, double access, big plot, and housing estate. Currently, most of the plots range from 600 to 900 m² in area, much larger than the latest provincial standard (350m²). As the land-use regulation relaxes in the rural housing area, bigger plots may be able to accommodate mix-use programs.

耕地资源
Farm Resource

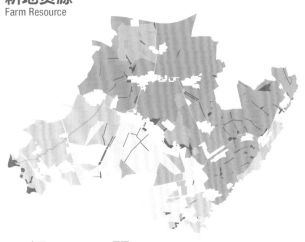

水田 Rice Paddie   果园 Orchard
菜田 Vegetables   花田 Flower Field
旱田 Dry Field   棚室 Greenhouse
树 Tree   养殖场 Animal Farm
试验田 Research   水体 Water Body

人口密度
Household Density

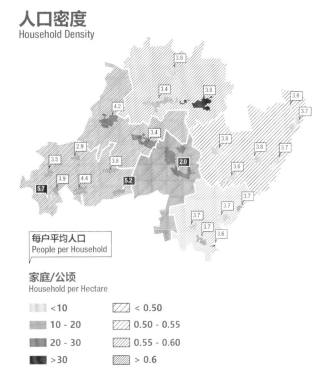

每户平均人口
People per Household

家庭/公顷
Household per Hectare

<10	< 0.50
10 - 20	0.50 - 0.55
20 - 30	0.55 - 0.60
>30	> 0.6

建筑密度
Building Density

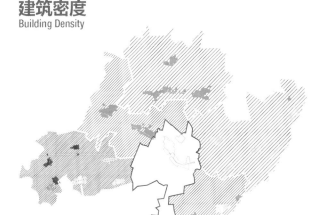

耕地占比
% of Farm land

100

0

耕地与建筑用地之比
Farm Land to Buildable Land Ratio

< 10
10 - 20
>20

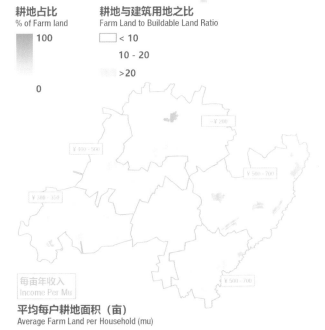

每亩年收入
Income Per Mu

平均每户耕地面积（亩）
Average Farm Land per Household (mu)

<10
10 - 20
20 - 30
>30

建筑面积/公顷
Building Area (m²) per Hectare

N/a	
<1000	< 60
1000 -2000	60 - 80
>2000	> 80

东兴村
Dongxing Village

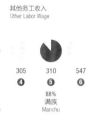

矿泉
Mineral Spring

龙甩湾采摘园
Longshuaiwan
Agritainment

总面积
Total Area — 1110 hm²

832 hm² 耕地 Farm Land	57 hm² 宅基地 Home Area	11 hm² 企业 Business

人均收入
Income Per Capita — 17916 CNY

~3300
农业种植收入
Farming Income

其他务工收入
Other Labor Wage

总人口
Total Population — 1888

367 未成年 Underage (<18)	532 壮年 Adult (18 - 45)	506 中年 Late Adult (45 - 60)	479 老年 Senior (60+)

62 ❶	441 ❷	223 ❸	305 ❹	310 ❺	547 ❻

2%
满族
Manchu

95%
满族
Manchu

88%
满族
Manchu

农业概览
Agriculture

玉米 Corn
基本作物
Primary Produce

榛子 Hazelnut
经济作物
High Value Produce

棚室 Greenhouse
适合农业项目
Farming Option

果园 Orchid
适合农业项目
Farming Option

大田 Open Field
适合农业项目
Farming Option

当地企业
Local Business

× 9
轻制造业
Light Manufacture

× 6
家庭作坊
Family Workshop

× 2
农业合作社
Agricultural Cooperative

东平村
Dongping Village

总面积
Total Area — 510 hm²

317 hm² 耕地 Farm Land	26 hm² 宅基地 Home Area	6.5 hm² 企业 Enterprise

人均收入
Income Per Capita — 17635 CNY

~2640
农业种植收入
Farming Income

其他务工收入
Other Labor Wage

总人口
Total Population — 1072

288 未成年 Underage (<18)	372 壮年 Adult (18 - 45)	197 中年 Late Adult (45 - 60)	215 老年 Senior (60+)

388 ❶	236 ❷	206 ❸	114 ❹	128 ❺

79%
满族
Manchu

农业概览
Agriculture

水稻 Rice
基本作物
Primary Produce

蔬菜 Vegetable
经济作物
High Value Produce

棚室 Greenhouse
适合农业项目
Farming Option

大田 Open Field
适合农业项目
Farming Option

当地企业
Local Business

× 3
轻制造业
Light Manufacture

× 1
食品加工
Food Processing

× 1
农家乐
Agritainment

× 1
度假村
Resort

东胜村
Dongsheng Village

东北农业大学
试验基地
Northeastern Agriculture
Institute Experiment Field

总面积
Total Area — 1113 hm²

675 hm² 耕地 Farm Land	44 hm² 宅基地 Home Area	7 hm² 企业 Business

人均收入
Income Per Capita — 17000 CNY

~980
农业种植净收入
Farming Net Income

其他务工收入
Other Labor Wage

总人口
Total Population — 2280

290 未成年 Underage (<18)	758 壮年 Adult (18 - 45)	763 中年 Late Adult (45 - 60)	469 老年 Senior (60+)

958 ❶	400 ❷	696 ❸

农业概览
Agriculture

玉米 Corn
基本作物
Primary Produce

榛子 Hazelnut
经济作物
High Value Produce

牛 Cow
规模养殖
Live Stock

果园 Orchid
适合农业项目
Farming Option

大田 Open Field
适合农业项目
Farming Option

棚室 Greenhouse
适合农业项目
Farming Option

当地企业
Local Business

× 3
轻制造业
Light Manufacture

× 1
食品加工
Food Processing

× 2
养殖场
Animal Farm

× 2
农业合作社
Agricultural Cooperative

× 3
农业供应
Farming Supply

向阳村
Xiangyang Village

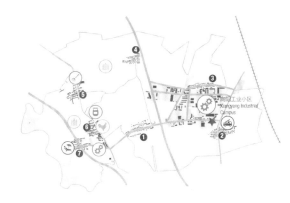

总面积 Total Area	890 hm²

518 hm² 耕地 Farm Land	40 hm² 宅基地 Home Area	120 hm² 企业 Business

人均收入 Income Per Capita	17000 CNY

~1200 农业种植收入 Farming Income	其他务工收入 Other Labor Wage

总人口 Total Population — 1979

370 未成年 Underage (<18)	546 壮年 Adult (18 - 45)	603 中年 Late Adult (45 - 60)	360 老年 Senior (60+)

314 ❶	393 ❷	277 ❸	167 ❹	262 ❺	280 ❻	286 ❼

农业概览 Agriculture
土地沙层多无法存水　无法做水田
Some soil is too sandy to be nice paddies.

玉米 Corn 基本作物 Primary Produce	水稻 Rice 基本作物 Primary Produce
棚室 Greenhouse 适合农业项目 Farming Option	果园 Orchid 适合农业项目 Farming Option

当地企业 Local Business

⚙ x 26 轻制造业 Light Manufacture	x 4 食品加工 Food Processing

石槽村
Shicao Village

总面积 Total Area	1070 hm²

550 hm² 耕地 Farm Land	21 hm² 宅基地 Home Area	35 hm² 企业 Business

人均收入 Income Per Capita	17900 CNY

~2000 农业种植收入 Farming Income	其他务工收入 Other Labor Wage

总人口 Total Population — 2053

292 未成年 Underage (<18)	766 壮年 Adult (18 - 45)	555 中年 Late Adult (45 - 60)	440 老年 Senior (60+)

840 ❶	1213 ❷

农业概览 Agriculture

玉米 Corn 基本作物 Primary Produce	蔬菜 Vegetable 经济作物 High Value Produce

当地企业 Local Business

⚙ x 5 轻制造业 Light Manufacture	x 3 食品加工 Food Processing

东方红社区
Dongfanghong Community

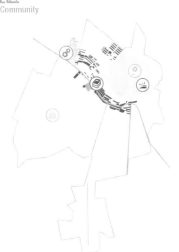

总面积 Total Area	990 hm²

381 hm² 耕地 Farm Land	27 hm² 住宅区 Residential Area

人均收入 Income Per Capita	

~2000 农业种植收入 Farming Income	其他务工收入 Other Labor Wage

总人口 Total Population — 1278

141 未成年 Underage (<18)	406 壮年 Adult (18 - 45)	379 中年 Late Adult (45 - 60)	352 老年 Senior (60+)

农业概览 Agriculture

水稻 Rice 基本作物 Primary Produce	薰衣草 Lavender 经济作物 Primary Produce
梅果 Berries 经济作物 High Value Produce	

当地企业 Local Business

⚙ x 1 轻制造业 Light Manufacture	x 1 主题种植 Themed Plantation
x 1 度假村 Resort	

李太屯 LITAITUN

西齐家屯 XIQIJIATUN

山根屯 SHANGENTUN

于始良屯 YUSHILIANGTUN

刘二转屯 LIUERZHUANTUN

刘世明屯 LIUSHIMINGTUN

赵家油坊屯 ZHAOJIAYOUFANGTUN

张家油坊屯(东胜村) ZHANGJIAYOUFANGTUN

东方红社区 DONGFANGHONG

贺家屯 HEJIATUN

山湾屯 SHANWANTUN

徐家油坊屯 XUJIAYOUFANGTUN

四间房屯 SIJIANFANGTUN

永生屯 YONGSHENGTUN

雷家屯 LEIJIATUN

刘家店屯 LIUJIADIANTUN

前关家屯 QIANGUANJIATUN

杨洪业屯 YANGHONGYETUN

永合屯 YONGHETUN

代家屯 DAIJIATUN

张家油坊屯(向阳村) ZHANGJIAYOUFANGTUN

石橱屯 SHICAOTUN

唐文公屯 TANGWENGONGTUN

后关家屯 HOUGUANJIATUN

李太屯 LITAITUN

西齐家屯 XIQIJIATUN

山根屯 SHANGENTUN

于始良屯 YUSHILIANGTUN

刘二转屯 LIUERZHUANTUN

刘世明屯 LIUSHIMINGTUN

赵家油坊屯 HAOJIAYOUFANGTUN

张家油坊屯(东胜村)ZHANGJIAYOUFANGTUN

东方红社区 DONGFANGHONG

贺家屯 HEJIATUN

山湾屯 SHANWANTUN

徐家油坊屯 XUJIAYOUFANGTUN

四间房屯 SIJIANFANGTUN

永生屯 YONGSHENGTUN

雷家屯 LEIJIATUN

刘家店屯 LIUJIADIANTUN

前关家屯 QIANGUANJIATUN

杨洪业屯 YANGHONGYETUN

永合屯 YONGHETUN

代家屯 DAIJIATUN

张家油坊屯 (向阳村) ZHANGJIAYOUFANGTUN

石槽屯 SHICAOTUN

唐文公屯 TANGWENGONGTUN

后关家屯 HOUGUANJIATUN

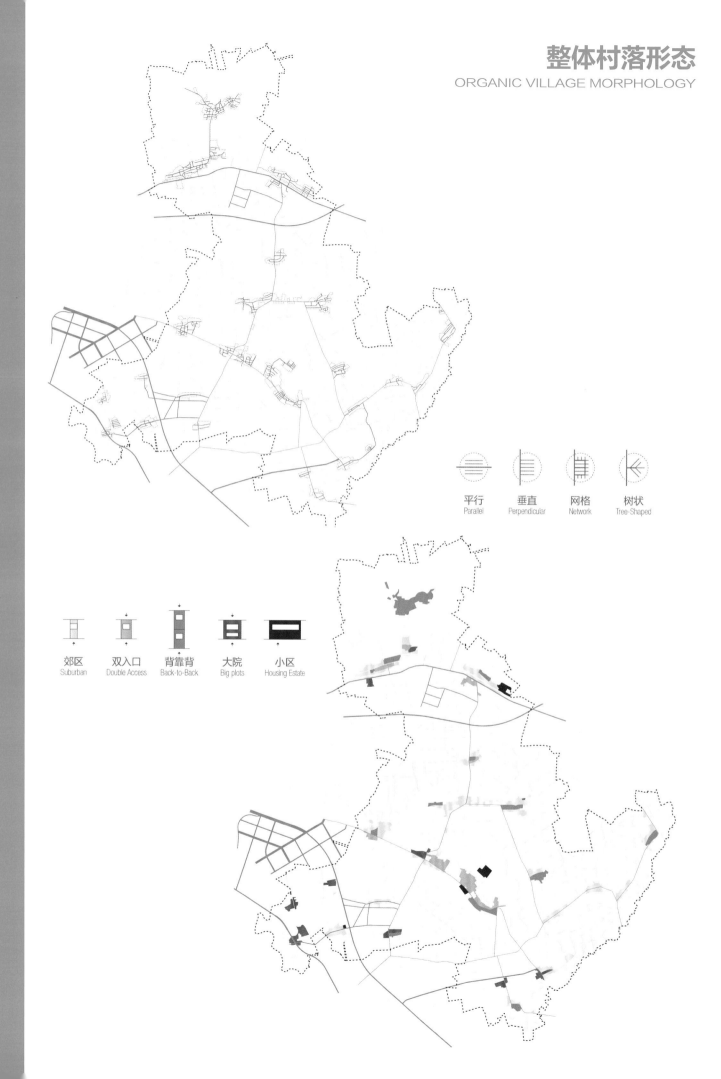

平行
Parallel

垂直
Perpendicular

网格
Network

树状
Tree-Shaped

郊区
Suburban

双入口
Double Access

背靠背
Back-to-Back

大院
Big plots

小区
Housing Estate

平行
Parallel

垂直
Perpendicular

树
Tree

网格
Grid

地形
Topography

结构
Structure

土地覆盖物
Land Cover

航拍
Aerial

阿什河 Ashi River

杨洪业灌渠 Yanghongye Irrigation Channel

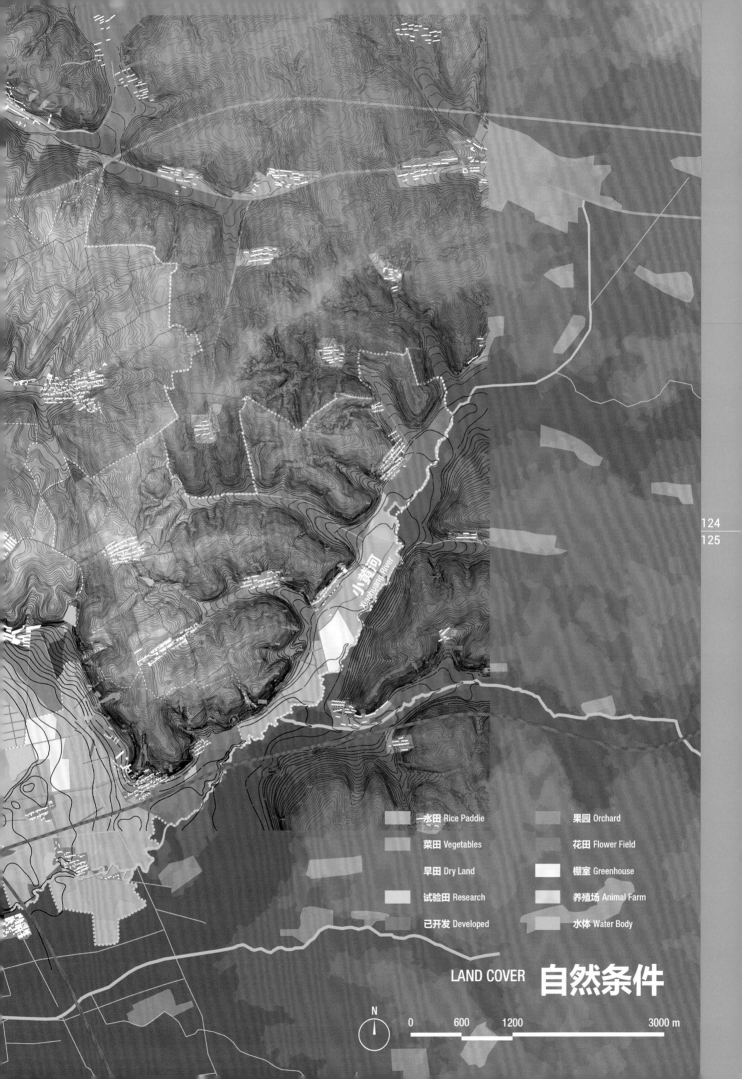

小黄河
Xiaohuang River

水田 Rice Paddie   果园 Orchard

菜田 Vegetables   花田 Flower Field

旱田 Dry Land   棚室 Greenhouse

试验田 Research   养殖场 Animal Farm

已开发 Developed   水体 Water Body

LAND COVER **自然条件**

N

0   600   1200   3000 m

哈尔滨绕城高速
Harbin Outer Loop Highway

长江路
Changjiang Road

阿什河 Ashi River

石槽村
Shicao Village

新香坊北站
New Xiangfang North

向阳村
Xiangyang Village

东方红社区
Dongfanghong
Community

江南中环路
Jiangnan Zhonghuan Road

哈阿公路
Harbin-Acheng R

佳哈高速
Jiamusi-Harbin Expressway

哈同公路
Harbin Tongjiang Road

东胜村
Dongsheng Village

东兴村
Dongsheng Village

哈同公路

东平村
Dongping Village

哈牡高铁
Harbin-Mudanjiang High Speed Rai

	住宅 Residential		公共绿地 Public Park
	二类工业 Type II Industrial		仓储 Warehouse
	教育研究 Education		市政设施 Utility
	重点开发 Special		其他农林用地 Other Agriculture
	卫生所 Health Care		基本农田 Core Farm Land
	政府 Goverment		水体 Water Body

LAND USE 土地用途

N

0   600   1200   3000 m

哈尔滨绕城高速
Harbin Outer Loop Highway

长江路
Changjiang Road

阿什河 Ashi River

江南中环路
Jiangnan Zhonghuan Road

新香坊北站
New Xiangfagn North

哈阿公路
Harbin-Acheng R

成高子站
Chenggaozi Station

佳哈高速
Jiamusi -Harbin Expressway

哈同公路
Harbin-Tongjiang Road

宾城铁路
BinCheng Rail

哈牡高铁
Harbin-Mudanjiang High Speed Rail

TRANSPORTATION
NETWORK 交通网络

N

0   600   1200   3000 m

在向阳镇选择三个重要节点进行城市设计，以展示如何开发和实施城市项目。它显示了另一种规模，其中农田、新经济节点、市民轴线和城市架构与当前村庄及其扩展区相匹配。

通过这些节点设计，来阐述整个地区是如何开发的，整体概念能够确保各部分最终整合效果，要比简单的拼凑更好。

这些放大的节点设计，可用来证明项目功能的丰富性，以及其对不同建筑形态的适应性。开放空间也被认为是开发成功的关键，这一点我们可以通过引入许多参考项目来说明。

This part of the Project is concentrated in three examples in Xiangyang to show how the Urban Project can be developed and implemented. It shows another scale where agriculture coverage, new economic nodes, civic axis and urban armature are matching together with current villages and their expansions.

It appears quite clear the need for some "pilot projects" that can be demonstrative of how the Overall Project can be developed, keeping in mind that Overall concept is the one that can ensure that the sum of the parts can be much better that the simple addition of the single pieces.

Zooms to test the different sectors.

Within the project several zooms of the some urban sector are presented to proof the richness of the programs, but also the adaptability of the different morphologies proposed on it. Open spaces are also considered key for the success of the development and this is the reason that many references are brought in the project.

新的城市区域将会找到与农村和谐共处并提高其品质的方式
NEW CITY EXTENSIONS SHOULD FIND THE WAY TO "WORK" WITH AGRICULTURAL SETTLEMENTS AND ENHANCE THEIR QUALITIES

薰衣草公园、温泉城和民俗风情园连接起来

JOINING LAVENDER PARK WITH SPA CITY
AND ETHNIC AGRITAINMENT

将功能节点与现有功能结合
CONSOLIDATING NODE WITH EXISTING USES

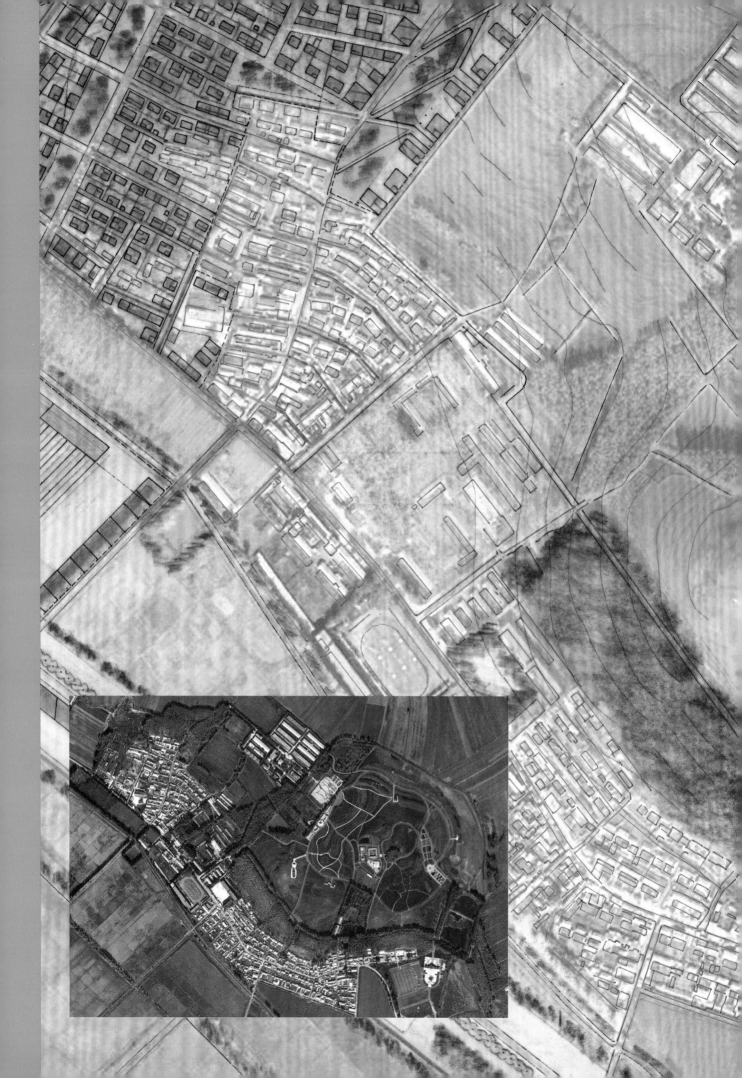

细部试验
DESIGN TEST

　　根据此前初步的分区试验和元素构成试验，我们便可以聚焦到某一个具体区域以完成更为精确的模拟提案。在这个过程中，我们将会考虑所有相关的城乡实地条件，包括地形、灌溉、基础设施、城乡形态、路网等。因此，该试验将会引申出更为准确的用地分界和适用功能。这也为具体的专业性的农业可行性研究奠定了基础。

Based on the results of the preliminary zoning and block configuration test, we will focus on a key area for more precise simulation. In the process, we will factor in all relevant urban and rural conditions on site, including topography, irrigation, infrastructures, morphology, road network, etc. The test will yield more precise zoning division and applicable programs, which lays the ground for detailed professional feasibility studies.

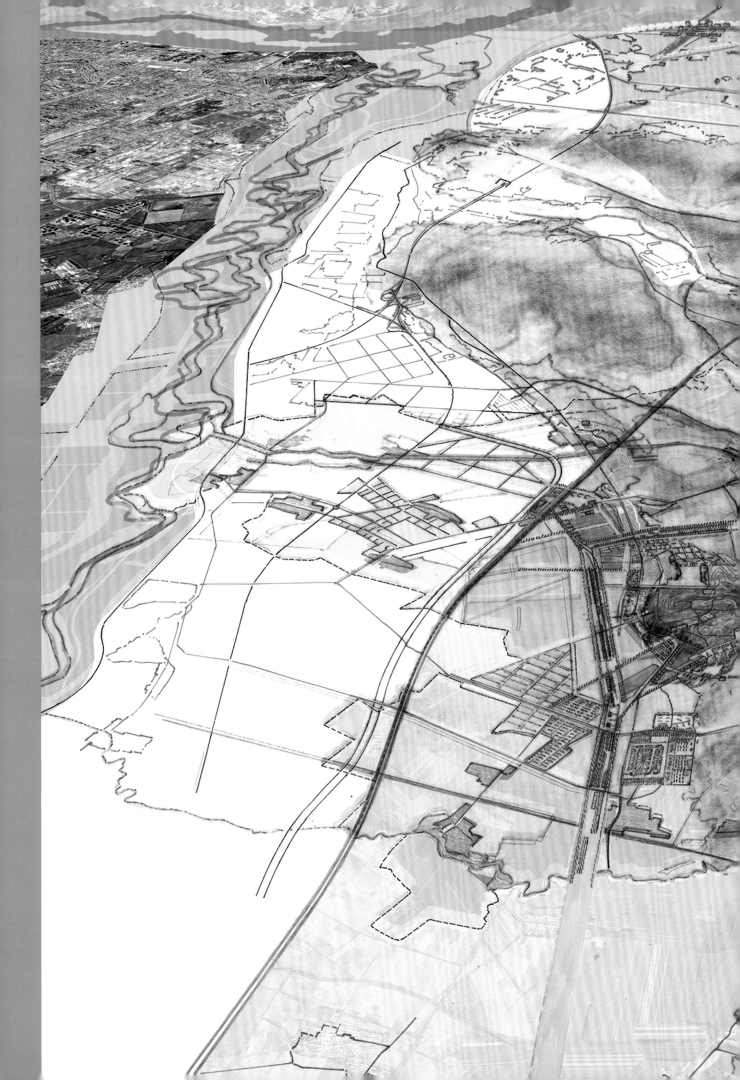

公众轴线为现有及规划中的节点带来价值

CIVIC AXIS PUT VALUE IN EXISTING AND PROPOSED NODES

运营管理策略

MANAGEMENT AGENCY

FOR XIANGYANG AND LARGE PARK

REGIONAL AGRICULTURE

INFLUENCE

我们可以列举许多案例来表明城市与乡村之间的积极对话或与城市相关的农业环境有多么重要。仅举几例：

米兰郊区农业园（意大利，从 2009 年开始）。这是一个非常大的农业园区，耕地占地面积在 47000hm^2 以上；主要目标是在激活复兴该区域之前将土地保护作为首要行动。

戴尔略弗雷特三角洲（西班牙巴塞罗那，2010 年）。这项规划占地 3000hm^2，毗邻国际机场，位于巴塞罗那的大都市周边。主要目标是保护丰富的农业形式免受大都市增长破坏。

谢尔比农场（孟菲斯，美国，自 2007 年以来）。被遗弃的旧址正在向一系列与自然景观相容的"景点"转型。其中有许多休闲活动。

伊夫林公园（巴黎，"绿色三角"，2016 年）。在大巴黎附近的萨克莱试图在许多经济活动即将发生的区段建立动态农业用途。公园还考虑了与农村用途兼容的新型低层住宅。

米兰郊区农业园，意大利，从2009年开始。
Parco Agricolo Milan, Italy, from 2009.

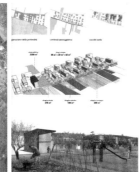

伊夫林公园，巴黎，"绿色三角"，2016年。
Yvelines Park, Paris, "Triangle Vert", 2016.

在建筑和变革领域推广新范式的展览
EXHIBITIONS TO PROMOTE NEW PARADIGMS IN ARCHITECTURE AND TRANSFORMATIVE TERRITORIES

戴尔略弗雷特三角洲，西班牙巴塞罗那，2010年。
Delta Llobregat, Barcelona, Spain, 2010.

Many are the examples we can point it out to show how important is the positive dialogue between city and countryside, or agricultural environment related to the city. Just to name few of them:

–Parco agricolo Milano, Italy, from 2009. It is a very large agricultural park, covering above 47000 hectares of cultivated land; main target is protection of the land as first action before managing the whole activation of it.

–Delta LLobregat, Barcelona, Spain, 2010. It is an initiative covering 3000 hectares next to the international Airport and within the metropolitan perimeter of Barcelona. Main objective is protecting rich agriculture form being devastated by the metropolitan growth.

–Shelbi Farm, Memphis, USA, since 2007. Abandoned site in process of transformation towards a series of "attractions" that are compatible with natural landscape. They include many leisure activities.

–Yvelines park, Paris, "Triangle vert", since 2016. Near Saclay in the Grand Paris influence trying to establish dynamic agricultural uses in the sector where many economic activities are going to happen. Park consider also new low-rise housing compatible with rural uses.

谢尔比农场，孟菲斯，美国，自2007年以来。
Shelbi Farm, Memphis, USA, since 2007.

IBA 即"国际建筑展览",是一种 100 多年前在德国创建的用来推广建筑的"特别展览"方法,但现在也是一种在荷兰和奥地利广泛使用的工具,可以促进许多不同的变革过程;国家在启动该倡议方面发挥了重要作用,同时许多利益相关方正在发起这项倡议。在我们的议题中,其中一些会很有参考价值:

① IBA 84-87。这是最知名的案例,它决定重建被二战破坏的柏林中心,从而对这个城市产生了巨大的影响,并为国家的统一创造了条件。它遵循 1957 年 IBA 开发的名为 Inter-Bau 的方式,产生了该市第一个纯粹的现代主义区。

②公园镇。加强现有经济以提高效率,这是发生在荷兰林堡的过程。汇集了 8 个小城市,以创造更强大的实体;始于 2017 年,并计划必须在 2020 年完成。

③埃姆舍公园。重新启动一个废弃的矿区,并使整个地区的经济向服务业转变。"国际建筑展览"是否能够通过将以前的废弃土地和工业融入新的用途和吸引力来创造新的景观形式?

④瑞士的巴塞尔 2020。它试图推动跨越国界的发展,"跨越国界,共同发展"是该项目的主题。

⑤维也纳 2022,奥地利。城市在社会住房经验方面拥有丰富的经验,试图接受并完成未来居住的新挑战。

⑥图林根 2023。斯塔德兰德的目的是同时在城市和农村集结行动。在经过一些提前的研究和市政当局的倡议之后,国家在 2012 年发布了第一项决议。

The IBA meaning "Internationale Bauausstellung" is a method of "special exhibition" created in Germany for more than 100 years to promote architecture but now is becoming a tool that is under use also in the Netherlands or in Austria to promote many different transformative processes; the State has a role launching the initiative, but many stakeholders are taking place on it. Some of them can be interesting in our discussion:

a. IBA 84-87. It is the most well know case. Decided to rebuild the center of Berlin still damaged by the WWII. It had a tremendous impact in the city and created the conditions for the reunification of the country. It follows the path of previous IBA develop in 1957 named Inter-Bau, that produced the first pure modernist quarter in the city.

b. Parkstad. Reinforcing existing economy to make it more efficient. This is the process in Limburg, The Netherlands; assembling together 8 small municipalities to create stronger entity; started in 2017 and must be completed in 2020.

c. Emscher Park. Reactivating an abandoned mining region and creating a change in the whole economy of the territory towards a service industry. Was an IBA able to create new forms of landscape by integrating former derelict lands and industry into new uses and attractors?

d. Basel 2020 in Switzerland trying to impulse the development of the sites across the borders. "growing together across borders" is the theme of the project.

e. Viena 2022, Austria. City with lots of experiences in social housing trying to investigate and promote new challenges for the living in the future.

f. Thüringen 2023. Stadland with the aim to assemble the action on the urban and the rural simultaneously. State took first resolution in 2012, after some previous research and initiatives by the municipalities assembled on it.

IBA 84-87

公园镇 / Parkstad

埃姆舍公园 / Emscher Park

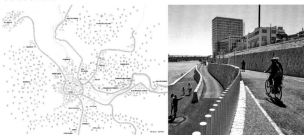
巴塞尔 2020 / Basel 2020

维也纳 2022 / Viena 2022

图林根 2023 / Thüringen 2023

城市农业是使我们的大城市更具弹性的主要战略。例如，2006 年伦敦市长通过的总体规划"伦敦粮食战略"对更好地利用农业开发，对城市和农村实体空间如何建立更好、更丰富的模式具有重要意义；尽管城市和农村被多次认为是互相矛盾的两个实体。

事实上，有一些一般性的步骤可以用来理解现在有关粮食和农业的讨论正在成为超越与农村相关的经济政策的内容，它已成为全面理解"战略规划和政策制定"学科的重要组成部分。

为了记住几个关键问题，我们必须考虑"罗马俱乐部"的重要性，这在 1968 年已经成为可持续世界讨论的基础。此外，澳大利亚于 2007 年由比尔·莫里森和大卫·福尔摩格雷姆进行的"永久性农业之一"的研究，也在寻找可持续的人类栖息地，并试图将农村改良和农业开发与包容性城市发展相结合。

在许多倡议中，我们应该提到 WTO（世界贸易组织），其使命是监督和改善不同国家之间的贸易，试图避免贸易中的食品捷径和投机举措。此外，自 20 世纪 20 年代在美国的农业生产中的创新举措"杂交种子"被研究和开发之后，"斯瓦尔巴特全球种子库"提出建立"种子库"作为传统基因遗产的倡议。

在这种背景下，我们可以看到许多大都市正在以不同的战略方式整合"城市与农业"。正如底特律试图利用大都市空间内的城市动态变化来重新引入农业产业作为空旷城市土地的经济驱动力的战略；或者是 2050 年东滩项目的经验，它位于上海附近崇明岛的一角，是将大型特大城市附近的开发与农业结合起来的一种方式。像柏林布里茨的"小型农园"这样的其他举措，对应于社区规模，为感兴趣的人群提供农业园林。我们还可以提到案例研究章节中的"垂直农业"例子，如纽约曼哈顿的阿尔博温室，基于水培生产的概念使生长强度翻倍。

Urban Agriculture is a major strategy to make our large cities more resilient. For example, the Masterplan of the Mayor of London in 2006 to take better advantage of the agricultural exploitation by the "London Food strategy" had important implication about the way cities and rural spaces can be establishing better and richer models for both entities; that for many times were presented as contradictorial entities.

In fact, there are certain general steps to understand how far today the discussion about the Food and Agriculture is becoming something that goes beyond the economic policies related to the rural world and it becomes an essential part of the overall comprehensive "strategic planning and policy making" disciplines.

Just to remember few key issues we must consider the importance of the "Club di Rome" in 1968 establishing already the basis for the sustainable world discussion. Also, the research done in Australia on "Permaculture One" by Bill Mollison and David Holmgrem in1978, searching for a sustainable human habitat, trying to integrate rural improvement and agricultural exploitation with inclusive urban development.

Among many initiatives we should mention WTO –World Trade Organization– with the mission to supervise and improve trade among different countries, trying to avoid the food shortcuts and speculative initiatives among the Trade. Also, the "Svalbard Global Seed Vault" as initiative to establish a "seed bank" as traditional genetic patrimony after the research and development od the "hybrid seeds" since 1920's in US, that had created an innovative process within agricultural production.

Within this context we can see the way of many Metropolis are addressing different strategies integrating "city and agriculture". Just as reminder the strategy of Detroit trying to use the change in urban dynamics within the metropolitan space to reintroduce agro-industry as economic devise for the empty urban land; or the experience of Dongtan Project for 2050, in the tip of the Chongming Island near Shanghai as a way of combining development and agriculture near that large megalopolis. Other initiatives like "Kleingärtens" in Berlin Britz, are corresponding to a more community-oriented scale, providing agricultural gardens for most of the interested population. We should also mention the "vertical agriculture" examples as referred in the chapter of the case studies, like Arbor House in Manhattan, NY, always based in the idea of hydroponic production multiplying the intensity of it.

很多城市正在使不同模式的聚居方式和谐共处
OTHER CITIES ARE MAKING THE DIFFERENT SETTLEMENT PATTERNS WORK TOGETHER

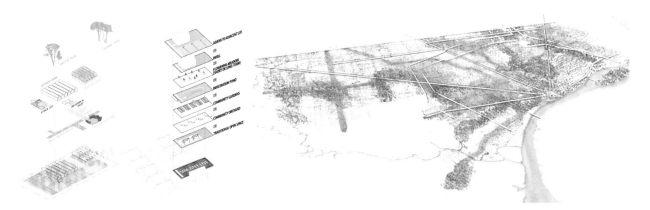

底特律重新引入农业产业作为经济驱动力　　Detroit Reintroduce Agro-industry as Economic Devise

2050东滩项目，崇明岛，上海　　2050 Dongtan Project, Chongming Island, Shanghai

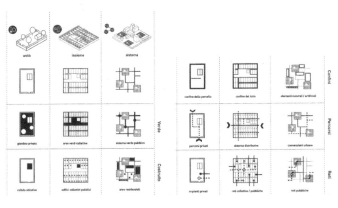

"小型农园"，柏林布里茨　　"Kleingärtens", Berlin Britz

纽约曼哈顿的阿尔博温室　　Arbor House in Manhattan, NY

关于实施的讨论需要清楚地了解不同机构和利益相关者的能力

THE DISCUSSION ON THE IMPLEMENTATION REQUIRES A CLEAR UNDERSTANDING OF THE CAPACITIES OF DIFFERENT INSTITUTIONS AND STAKEHOLDERS

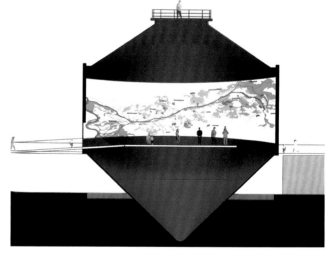

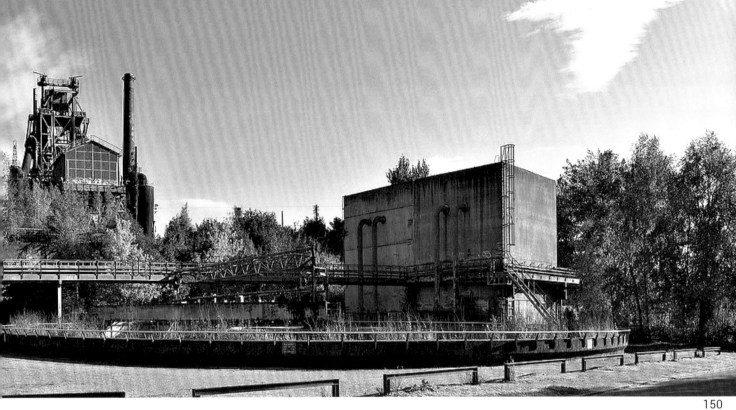

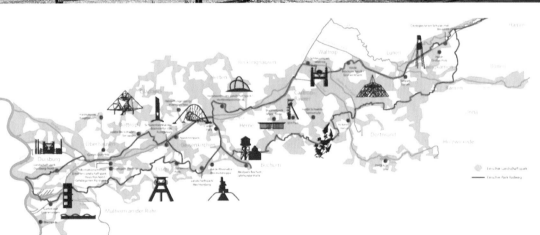

埃姆舍公园曾是一片荒废的工业用地。通过一点一滴的努力，15年后，这里成了欧洲最为成功的地区之一。

EMSCHER PARK WAS A DERELICT INDUSTRIAL LAND. PIECE BY PIECE, AND AFTER 15 YEARS, IT HAS BECOME A VERY SUCCESSFUL PLACE IN EUROPE

UCAP 创意长卷农业公园项目显示了很高的一致性，将不同的重要目标混合在一起：以农业产业作为主要驱动力；聚焦于经济发展；引入休闲和旅游以保证更高的服务水平；以及提高该区的村庄和新住宅的容量。

由于古老的农业传统和哈尔滨大都市的影响，UCAP 项目旨在改变向阳中心的概念结构，以应对其不确定性。在原农业区，建议实施面积超过 200km² 的景观规划和城市开发项目，以实现生态、经济和农业更新。在该地区，景观和城市设计首次为哈尔滨大都市的影响建立了新的区域特征。

这里介绍的项目建立在真实资产的基础上，以创建一个变革性的战略。我们的假设是将其定义为"特殊项目"，它需要一个特殊的实体或"机构"来进行开发。

根据"国际建筑展览"IBA 的多重经验，这需要一些机构支持。各方的广泛引导参与将使其成为可能：国家、省政府、哈尔滨市、向阳区和市民亟须知情并参与，以确保使其成为可能。

当确保政治支持后，许多不同的利益相关者——来自本地、国内和国际市场的投资者、开发商等——就可以轻松加入不同的项目中。

这里提出的项目可以作为实施 UCAP 创意长卷农业公园的结构计划；正如我们在其他例子中所看到的，它可能为大都市这一部分的区域经济发展创造自己的身份。在理想情况下，最好建立一个负责 UCAP 项目开发的"机构"。UCAP "机构"应协调不同地点的规划和项目开发行动。

每个机构都应根据自己在区域开发中的权利而负有相应责任。最终，国家可以提供"特殊"项目所需的条件和这个极端气候条件下农业产业创新项目所需的财政支持，项目能够产生改进并可作为适用于中国或国外其他背景的示范实验项目。最终这种财政帮助可以包括一些书面奖励或其他形式的激励。其他机构可以帮助开展"绿色基础设施"的"关键"投资和少数试点项目。也许区政府可以承担改善村庄的责任。

我们可以想象，向阳可以作为整体项目的第一步，可以作为"样本"或可以修改和调整的试验。然而，一旦项目处于共享和探讨的过程中，实施部分必然会被视为待讨论的开放环节，并根据时间表和待确定的优先事项进行重新安排。

The Project for the UCAP park shows a great level of consistency mixing together different important ambitions: From one side agro-industry as main driver; concentration in the impulsive of the economy; introduction of leisure and tourism to support another level of service; and the reinforcement of the villages and new residential capacity in the District.

Project for the UCAP is aiming to change the conceptual structure of the central Xiangyang at large, in response to its uncertainty because of old agricultural tradition and metropolitan influence from Harbin. In the former agricultural region, landscape planning and urban development projects covering over 200 square kilometers are proposed to be implemented for ecological, economic and agricultural renewal. For the first time in the region, landscape and urban design can come to the fore to establish a new regional identity into the metropolitan influence of Harbin.

Project presented here is build on real assets to create a transformative strategy. Our hypothesis will be to define it as "Special Project" requiring a special entity or "Agency" for its development.

According the IBA multiple experience, several institutional supports are required. It is convenient a broad engagement of leadership to make it possible: State, Metropolitan Government, City of Harbin, Xiangyang District, and municipalities must be informed and engaged to make it possible.

When political support is ensured many different stakeholders-investors, developers, etc. from local, national and international market- may join easily the different initiatives.

Project as it is presented here may act as Structural Plan for the implementation of the UCAP park; as we can see in other examples it may create the identity for the regional and economic development of this part of the Metropolis. In this respect ideally, it will advisable to create an "Agency" responsible for the development of the UCAP Project. UCAP "Agency" should coordinate actions of planning and project developments in the different sites.

Each Institution may have its own responsibility according to its own rights in the development of the territory. Eventually State can provide the condition of "special" project and the financial frame for this innovative project for agro-industry in extreme climate condition as experiment able to produce improvement that can be applicable to other context in China or abroad. This financial help can eventually include some text bonus or other forms of incentive. Other institutions can help in developing the "key" investments in "green infrastructure" and few pilot projects. Perhaps improvement of the villages can be responsible of the District.

We can very well imagine that Xiangyang can be the first step of the Overall Project and may serve as "sample" or test that can be modified and adjusted. Nevertheless, implementation must be seen as open chapter to be discussed and rearrange according to the time schedule and priorities to be establish once the Project is in the process of been shared and debated.

埃姆舍公园提供了新的活动，改善了当地居民的生活，并吸引新的人群前来生活工作

EMSCHER PARK PRODUCED NEW ACTIVITIES AND IMPROVED THE LIVE OF RESIDENTS AND NEW ONES CAME TO LIVE AND WORK THERE

滨江湿地景区
Binjiang Wetland Pa

哈尔滨市区 Harbin Urban Area

人口   4,219,500
Population

可支配收入   ¥ 33,190
Disposable income

人口密度   413 人/平方公里
Urban Density   people/ km²

哈尔滨西站
Harbinxi Station

HARBIN

CREATING AGRO–PARK IN

AGRICULTURAL REGION

创意长卷
农业公园解析

松花江
Songhua

白鱼泡湿地公园
Baiyupao Wetland Park

农科院试验田
Agriculture Institute
Experiment Field

天恒山风景区
Tianhengshan Landscape Zone

东北农业大学试验田
Northeastern Agriculture Univeristy
Experiment Field

新香坊北
Xinxiangfangbei

薰衣草公园
Lavender Park

海都国际温泉城
Haidu Spa City

天汇小镇文化中心
Tianhui Cultural Town
Cultural Center

伏尔加庄园
Vodka Garden

闽南国际商贸城
Minnan International
Trade City

大型创意长卷农业产业园内的新型产业功能节点

产业功能编组是功能节点的生成器，建立在两个维度之上：一个是上一章所提到的三种不同受众的功能类型，另一个则是当地城乡发展条件，包括地形、自然资源、建筑形态等。根据规划给出的空间框架，在选择项目开发地点的同时，该地点的城乡发展条件都将作为已知。结合染色法所确定的功能类型，则可生成一系列具有潜力的开发项目类型。之后寻找案例并深入研究其运作机制和功能表现，找到领域专家深入考察现场条件，确定最终的功能项目定位以及规模，我们形象地称之为"印章"。

编组

根据受众和城乡发展条件这两个限制条件可以生成很多意料之外情理之中的创新功能，开拓产业思路。如温泉与本地建设交叉产生的温泉果园系统，抑或是水塘之上无土栽培的浮床稻产业，休闲旅游当中低高度建筑群产生的庙宇酒店。我们从所有生成的功能项目当中选择了试验田、主题公园、运动场、采摘园、社区广场、水疗度假、温泉灌溉系统、浮床稻、种植林、农业学校、农庄旅游、温室农业、温室花园、社区中心、表演中心、批发市场、养老医疗等18个功能项目作为之后印章的主题。

开放式规划手法

传统产业功能规划往往在规划开始之初便限定了未来几十年产业区块的定位、大小以及经济指标。对于经济结构多变，发展速率不稳定的城市边缘区域来说，这种产业功能规划手法将会限制其发展的可能性，不能最大化当地土地的利用价值，实现最佳的产业布局。甚至会造成资源的浪费，同时造就一系列与周边不协调的城乡形态，而荒废位于其中的土地。因此，我们在功能排布这一章中，给出了一种新型的、开放性的产业功能规划手法。首先我们建议构造一个概括性的创意农业公园（在之后的章节将作为创意长卷公园具体介绍），用以利用向阳镇内到松花江边可用的土地。根据当地适合发展产业的地块区位，利用三色定点染色法原理，根据某一时期的城市发展形势，确定三种不同受众的功能类型——农业产业、休闲旅游、本地建设——之间的协同关系。结合产业功能编组给出的框架结构和实例研究，确定该时期合适的产业功能定位和规模。开放性的规划手法是世界规划的一个新潮流。目前荷兰阿尔米尔市也在城市东部的农业地区进行开放性规划的试验。

印章

我们根据之前选定的功能项目寻找合适的本地和国际案例，尽可能寻找多功能、绿色、社会影响力广以及经济上可持续的案例。农业产业类的案例以研究其运营模式和盈利表现为主。休闲娱乐类的案例以研究游客来源和收入为主。本地建设类的案例以研究其设计和社会影响为主。

The nodes of the new economy are placed within the large UCAP agro-industry park

The matrix is a generator of programs for the nodes of economy, based on two dimensions: one is the three functional types based on the types of recipients as we mentioned in the last chapter; another is the local urban/rural conditions, including topography, natural resources, building morphologies, etc. When we zoom into one developable spot in our structural plan, its urban/rural conditions are fixed. Once we determines the functional type from the graph coloring exercise, the crossover between the urban/rural conditions and the functional type will generate a series of potential programs applicable to the location. Afterwards, we will look for exemplary cases, study their operation and performance, and get specialists in the field to study the detailed site condition to formulate the final program and scale. The end result is what we called "stamps".

Matrix

Using clientele type and urban/rural conditions as constraining conditions can yield many surprising yet reasonable programs. It provides an efficient framework for brainstorming. For example, when hot spring crossovers with local settle, there can be irrigation system for local orchard. The aquaponics on a pond can lead to floating rice growing. Low-rise enclaves for leisure tourism may as well lead to temple hotel. Among all the programs generated, we selected 18 as the topics for the stamps, including: experimental field, theme park, sports field, harvesting garden, community square, spa resort, thermal orchard, floating rice, tree farm, agriculture academy, agritourism, greenhouse farming, winter garden, civic center, performance center, wholesale market, elderly care.

Open Planning Method

Traditional functional and economic planning often dictates the subject, scale, and economic metrics of development projects decades before the plan becomes fully realized. For metropolitan areas at the urban fringe where economic structure is changing and growth rate unstable, such a rigid planning methodology will limit the development potential of an area and prevent the best use of the land. It may result in a massive waste of resources and inconsistent urban form that creates lots of intermediate derelict land. Therefore, in the chapter on programming development, we propose an innovative open planning method. According to this methodology, we will first suggest to create an Overall Park for Creative Agriculture (later will be explained as UCAP) to be able to improve the exploitation of the available fields in Xiangyang but till the Songhua River; second identify suitable locations for new developments, employ the 3-color vertex coloring based on graph theory, analyze a given period's economic development trend, and coordinate the synergy between three types of functions: agro-industry, leisure/tourism, local settlement. Combining with the framework and case study methodology in the economic and functional matrix chapter, we will determine suitable development subjects and their scales in that period. Open planning method is a new global trend of planning. Currently, Almere in the Netherlands is testing open planning to develop the agricultural area in the east of the city.

Stamps

Based on previously selected programs, we select local and international cases as our stamps. We have preference for mix-use, green, socially influential, and economically sustainable cases. We have different research priorities for different functional types. For agro-industry, we study their business model and financial performance. For leisure tourism, we focus on the visitor source and the income sources. For local settlements, we look into its design and social impact.

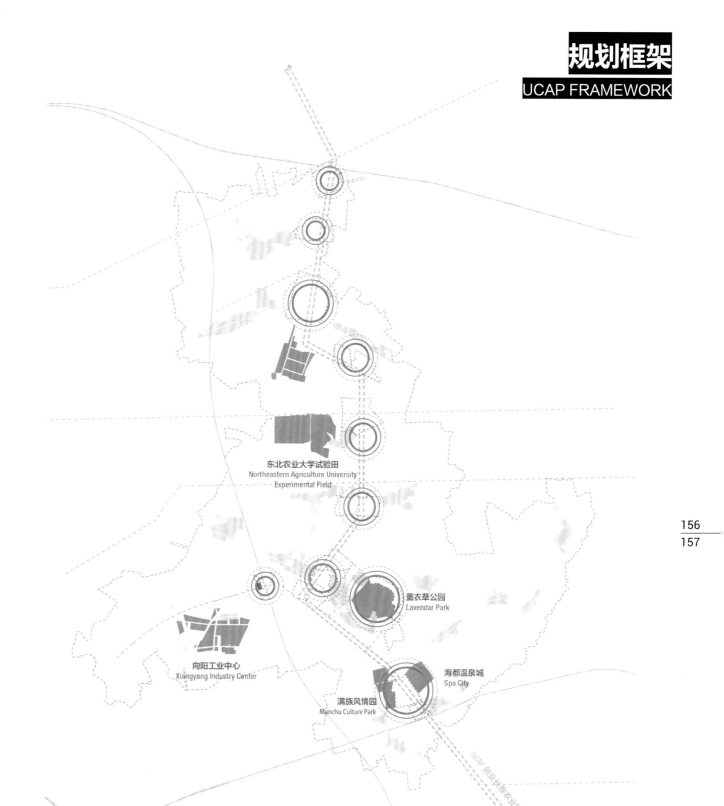

东北农业大学试验田
Northeastern Agriculture University
Experimental Field

薰衣草公园
Lavendar Park

向阳工业中心
Xiangyang Industry Center

海都温泉城
Spa City

满族风情园
Manchu Culture Park

创意长卷公众轴线穿过向阳镇中心，连接了多个现有的产业节点，包括东北农业大学试验田、高铁站、薰衣草公园、海都温泉城、满族风情园等。沿着公众轴线交替分布着产业节点和农业景观带。如何结合现有功能节点，产生协同效应，是之后产业规划的一项主要内容。

The civic axis of UCAP passes through the middle of Xiangyang town, connecting several existing nodes, including Northeastern Agriculture University Experimental Field, High-speed Railway station, Lavendar Park, Spa City, Manchu Culture Park, etc. The structural scheme places functional nodes and scenic space along the axis alternatedly. How to utilize existing nodes to form synergy is one of the main tasks of our economic planning.

区域项目可以通过具体步骤实施。在这里，可以看到一个讨论提案，优先考虑与高铁相连的更容易开始开发的部分。

District project can be implemented by steps. Here you can see a proposal for discussion giving priority to the sectors more connected to the HST that are easier to start been developed.

发展阶段

项目组织的方式可能会出现不同的发展步骤。为了满足这个需求，很重要的是基础设施都必须被精心规划和保障，同时功能和建筑可以根据整个地区的需求和演变而被开发。

大的原则是，优先考虑大的功能片区以及从南到北的城市节点。每个部分的方案略有不同，但形态是和谐一体的。另一方面，旅游项目是相当具体地能够提供更好和更丰富的活动供游客选择，后文有整个项目不同的设施和活动。

Steps for development

Project is organized in a way the different steps for development may occur. For this is important that infrastructures have to be well planned and ensured, and program and architecture can be developed according demands and evolution of the whole site.

In principle steps are considering large sectors or urban nodes from the south to the north. Each sector has slightly different programs but morphologies are coherent. On the other side, tourism program is quite site specific to be able to provide better and richer options for the visitors, following different amenities and activities considered for the whole Project.

分期实施和自然动态

总的来说，总体规划必须将时间作为一部分来构思，即分期，这对于评估不同的策略以便能够实施使公园存活的关键动作，以及创建可以解释和展示我们希望在过程中期实现的目标和品质的必要条件十分重要。

在这方面，必须理解和谨记"自然元素"和城市形态演变相比具有不同的创造和再生的节奏，而且我们必须了解完成第一步和第二步的重要性：我们确实认为第一区域已经准备就绪，我们希望看看公园的另一端——即西部，河流旁边——是否可以在第二阶段转换，通过在两端集中活动，来最大化公园的存在感。

Phasing and natural dynamics

In general Master Plan must be conceived in a way that time is part of it, and this is important to evaluate different strategies to be able to implement the key actions that make park already exists, and to create the minimum conditions that can explain and show the ambitions and the qualities we want to achieve in the midterm run.

In this respect is essential to understand and remind that "natural elements" have a different rhythms of creation and reproduction that urban form and city forms, and then we have to understand how important will be to fix the first and second steps: We do really feel first area is quite ready for action, we would like to see if the other end of the park -meaning the western part, next to the River- can be converted in a second phases, concentrating the actions in both ends, to maximize the presence of the park.

规划期限

规划应该试图合并和完成规划节点和现有节点。绿轴将连接那些节点直到创意农业园区沿着哈尔滨东完全展开。

到 2020 年，初始节点应该在开发中，不一定全部完成，这样以便为次要用途发挥提供空间并改善整体效果。

到 2025 年，节点的联系和协同作用应该开始沿着城市绿轴和在向阳内部展现。

2030 年及以后，"长卷"可以延伸到东哈尔滨。

Planning period

Planning should look for the consolidation and completion of both proposed and existing nodes. Civic axis will connect those till the Creative Agriculture Park is unrolled along east Harbin.

By 2020, initial nodes should be under development, not necessarily finish in order to give room for minor uses to take place and improve the whole.

By 2025, their connections and synergies should start taking place along the Civic Green Axis and inside Xiangyang.

2030 and beyond, the "unrolling" can be extended across east Harbin.

区域项目可以通过具体步骤实施
DISTRICT PROJECT CAN BE IMPLEMENTED BY STEPS

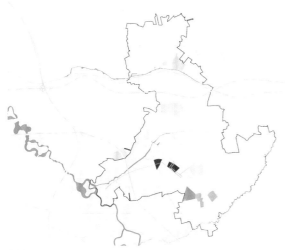

2018~2020
阶段1 整合哈牡高铁区域
PHASE 1 CONSOLIDATING HST SITE

2018~2020
阶段2 向现有节点延伸
PHASE 2 TOWARDS EXISTING NODES

2018~2020
阶段3 新的战略节点
PHASE 3 NEW STRATEGIC NODE

2020~2025
阶段4 向北连接各个节点
PHASE 4 LINKING NODES TOWARDS THE NORTH

2020~2025
阶段5 建立整个向阳的主轴
PHASE 5 SPINE FOR WHOLE XIANGYANG

2025~2030
阶段6 展开创意长卷农业公园
PHASE 6 UNROLLING THE CREATIVE
AGRICULTURAL PARK

创造并维持新开发项目之间的合力
PRODUCE AND MAINTAIN SYNERGIES CREATED BY EACH NEW DEVELOPMENT

农业产业
Agro-Industry

Experiment Field

浮床稻
Floating Aquaponics

稻田鱼
Rice Paddy Fish

休闲旅游
Leisure/Tourism

主题公园
Theme Park

水疗度假
Spa Resort

有机农场

采摘园
Harvesting Garden

温泉灌溉系统
Thermal Irrigation

当地村落
Local Settlement

社区广场
Community Square

温室农业
Intensive Farming

种植林
Tree Farm

批发市场
Wholesale Market

工业厂区
Industrial Campus

农庄旅游
Agritourismo

庙宇酒店
Temple Hotel

表演中心
Performance Center

游客中心
Visitor Center

商业街
Commercial Street

温室花园
Winter Garden

农耕文化园
Agri-Culture Park

社区中心
Civic Center

养老医疗
Elderly Care

对于需要强调的方面进行积极的管理
ACTIVE MANAGEMENT TO SUGGEST WHICH
BRANCH MUST BE ENHANCED

产业案例
ECONOMIC EXAMPLES

农业产业
Agro-Industry

东北农业大学试验田
向阳镇
Northeastern Agriculture University Experiment Field
Xiangyang Town

万州鱼种站汪家坝基地
重庆
Wanzhou Fishing Station Wangjiaba Base
Chongqing

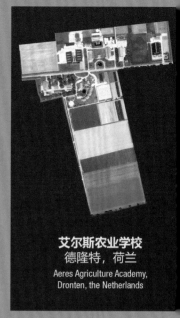

艾尔斯农业学校
德隆特，荷兰
Aeres Agriculture Academy,
Dronten, the Netherlands

休闲旅游
Leisure / Tourism

普罗旺斯薰衣草庄园
向阳镇
Provence Lavender Park
Xiangyang Town

满族风情园
向阳镇
Manchu Culture Park
Xiangyang Town

海都国际温泉城
向阳镇
Haidu International Spa City
Xiangyang Town

当地村落
Local Settlement

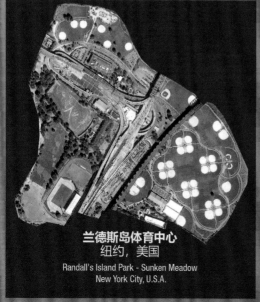

兰德斯岛体育中心
纽约，美国
Randall's Island Park - Sunken Meadow
New York City, U.S.A.

塔皮斯路广场
家乐福弗耶，海地
Tapis Rouge Square
Carrefour-Feuilles, Haiti

温泉果园灌溉系统
巴塞罗那，西班牙
The Irrigation System At The Thermal Orchards
Barcelona, Spain

印章
STAMPS

农业产业
Agro-Industry

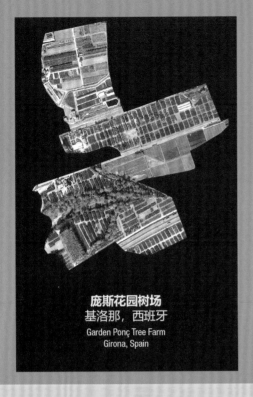

庞斯花园树场
基洛那，西班牙
Garden Ponç Tree Farm
Girona, Spain

沃星青椒基地
韦斯特兰，荷兰
Gebr. Valstar Green Peper Base
Westland, the Netherlands

新考文特花园市场
伦敦，英国
New Convenant Garden Market
London, UK

休闲旅游
Leisure / Tourism

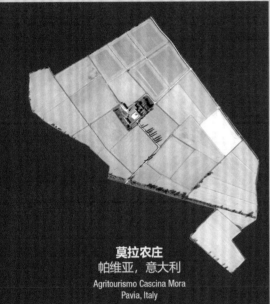

莫拉农庄
帕维亚，意大利
Agritourismo Cascina Mora
Pavia, Italy

米勒户外剧场
休斯顿，美国
Miller Outdoor Theater
Houston, U.S.A.

东景缘智珠寺酒店
北京
The Temple Hotel
Beijing

当地村落
Local Settlement

菲普斯植物园
匹兹堡，美国
Phipps Conservatory and Botanical Gardens
Pittsburgh, U.S.A.

市镇中心馆和广场
普罗旺斯，法国
City Center Pavilion & Square
Provence, France

米得东景老年社区
米得布里，美国
East View at Middlebury
Middlebury, U.S.A.

农贸批发市场——本地竞争
WHOLESALE MARKET — LOCAL COMPETITOR

① 哈达蔬菜水果批发市场 Hada Vegatable & Fruit Market

建造年份: Year Built	2002
占地面积: Land Area	14.1hm²
建筑面积: Building Area	240,000m²
总投资: Investment	¥ 500,000,000
交易量: Annual Tonnage	500,000
交易额: Annual Turnover	¥ 3,000,000,000
占地比率: Coverage:	61%
市场主体: Operation Entity:	私营部门 Private
买家范围: Clientele Range:	都市级 Metropolitan

② 哈尔滨花卉大市场 Harbin Flower Market

建造年份: Year Built	2015
占地面积: Land Area	10hm²
建筑面积: Building Area	30,000m²
规模: Scale	320 展位 shops
占地比率: Coverage	45%
市场主体: Operation Entity:	公共部门 Public
买家范围: Clientele Range:	都市级 Metropolitan

农业技术学院
AGRICULTURAL TRAINING
ACADEMY

2

1
3

1 黑龙江省农业技术推广站
Heilongjiang Agri-Technology Promotion Station

2 哈尔滨市农业技术推广服务中心
Harbin Agri-Technology Promotion Center

3 香坊区农业技术推广站
Xiangfang Agri-Technology Promotion Station

开放式规划手法

仿真模拟测试是开放式规划手法从概念走向现实的关键一环。如果说之前提出的结构性框架为预测未来的发展提供好了算法架构，仿真模拟测试便是根据意向规划指标，包括用地面积、功能类型、密度等，使用结构性框架进行模拟，预测相应的城市发展指标，如投资额度、新增居民数量、新增就业、人均收入等。这一过程中，规划人员可以根据城市阶段性发展的要求，使用仿真模拟测试的预测数值进行参考，不断地优化规划细节。同时，规划人员可以根据城市当时的经济发展标准，随时调整仿真模拟测试当中的控制参数，使测试结果能够尽可能接近现实。这种非线性的规划逻辑与传统的推导式规划逻辑相比，有着很高的灵活度，能够适应规划实践当中以十年计的时间跨度。

应用建议

在本章所展示的仿真测试当中示意的任何一个方案都有着深入设计的可能。同时，仿真测试不仅仅停留在大规模用地布局的规划，还可以用作分期规划甚至具体细节设计的依据，我们之后会专门对此进行讨论。目前来看，我们最初应该将精力集中到高铁站周边地区，完成 UCAP 第一个实践成果，而不是将关注点散落在区域各处而破坏整体项目的一致性。

Open Planning Method

Simulation is a key process in the open planning method where the structural scheme manifests itself in various versions and gets tested for performance. If we consider the structural scheme as the generator of new developments, the simulation process is to see what is generated under different conditions. The planner can take a variation of the structural scheme and use its key metrics—land budget, function type, density—to predict outcomes from the plan, including capital investment, new residents, new employment, and per capita income, etc. It is a useful tool for planers to optimize the details of planning and achieve the most current development goal. Moreover, planners can modify parameters in the prediction model according to the economic development in the city to make sure the simulation is updated and grounded in reality. Compared to the traditional linear logic of economic planning, this kind of non-linear planning logic is much more flexible and more likely to stay relevant over the entire implementation process of a master plan, which usually span across decades.

Recommendations

Any of these three scenarios or others that can be generated can be used for further development. Meanwhile, the application of simulation are not limited to large scale land use planning; it can also be used for phasing and detailed design, which we will explain in more details in the special chapter. It seems wise to concentrate first efforts near the new HST train and to the creation of one "real fragment" of the UCAP, rather than applying a more scattered distribution that will lose consistency for the overall project.

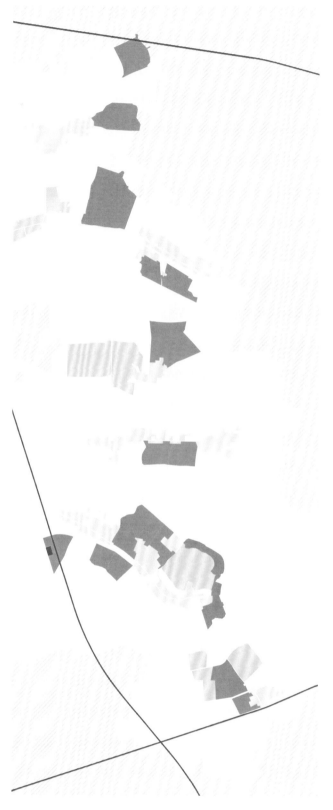

用地选择
PARCEL SELECTION
根据规划框架，选择用地范围
Select appropriate land parcels based on the framework

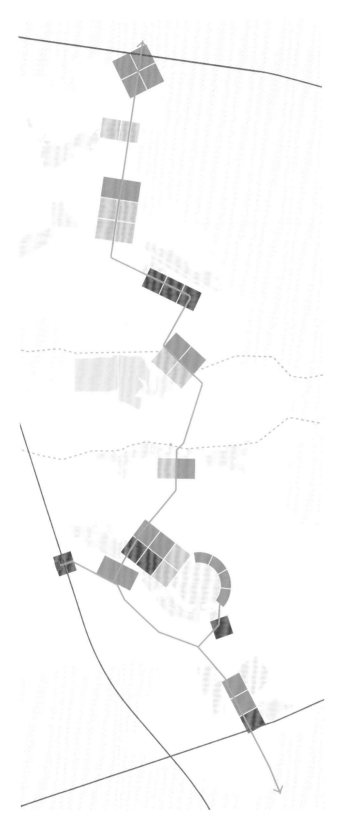

分区测试
ZONING SIMULATION

将用地初步分割并测试功能分区的经济表现
Divide the parcels and simulate zoning performance

实地测试
DESIGN SIMULATION

根据现实情况，测试具体形式及城乡效果
Simulate real design scenario based on real local conditions

功能节点

功能节点是产业布局的基础,有农业产业、休闲旅游和本地建设三大类。功能节点一般会有一定的建筑量。

景观空间

景观空间指的是高品质的农业景观带,但不一定以公共开放空间的形式。也可以是产业农庄内部的景观区,可供公众远观。

新设立的功能节点和景观带主要占用的是现有农田区域,包括一般农林用地和基本农田。其中功能节点占地约为497hm²,占全镇用地的7.2%,而景观带约为600hm²,占全镇用地的8.7%。考虑到基本农田保护的问题,红线范围内应当安排农业耕作相关的功能。

Functional Node

Functional Node is the basis of programing. There are three types of functions—agro-industry, leisure/tourism, and local settlement. Often functional nodes have certain volumes of buildings.

Scenic Space

Scenic space refers to high-quality agricultural landscape zone. It is available for the public to see, but it is not necessarily public open space. It could also be visible scenic areas within production farms.

Most of the new function nodes and scenic spaces occupy current farmland, both generic farmland and core ones. Function nodes cover 497hm², 7.2% of total town area. Scenic spaces cover 603hm², 8.7% of total town area. Condering the preservation of core farmland, the land within the red tape will be used for farming.

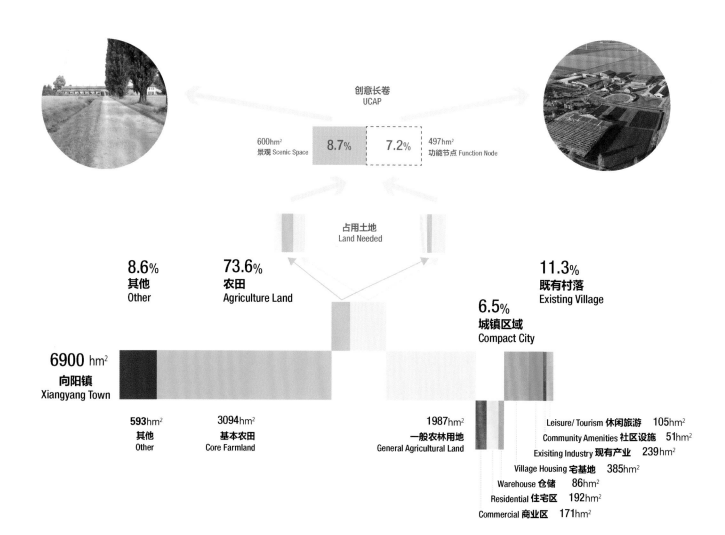

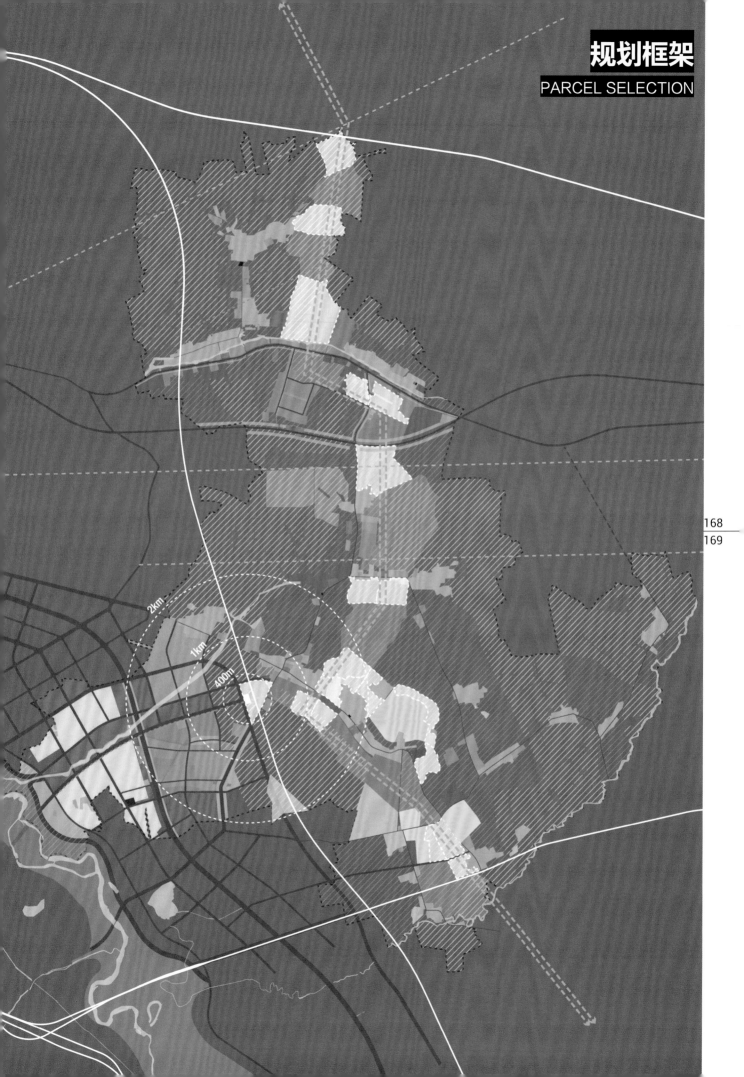

2km

1km

400m

假定条件 Assumptions

农业产业 / Agro-Industry

开发成本 / Development Cost (¥ /hm²)	¥9,000,000
利润率 / Profit Margin	15%
资产回报率 / Return on Asset	10%
营收 / Revenue (¥ /hm²)	¥6,000,000
单位面积劳动力 / Labor per area (hm²)	4
工资与净收入比值 / Wage to NOI Ratio	30%

休闲旅游 / Leisure Tourism

开发成本 / Development Cost (¥ /hm²)	¥3,000,000
利润率 / Profit Margin	15%
资产回报率 / Return on Asset	10%
营收 / Revenue (¥ /hm²)	¥2,000,000
单位面积劳动力 / Labor per area (hm²)	1
工资与净收入比值 / Wage to NOI Ratio	30%
游客人均消费 / Average tourist spending	¥50.00

本地建设 / Local Settlement

开发成本 / Development Cost (¥ /hm²)	¥1,000,000
住宅单元密度 / Density (unit/hm²)	25
平均月租 / Average Rent	800
营收 / Revenue (¥ /hm²)	¥240,000
就业密度 / Employment per area (hm²)	0.2
户均人口 / Average Household Size	2.4

向阳镇现状 / Xiangyang

人口 / Population	15809
人均收入 / Per Capita Income	¥17,000

#1 产业优先场景 Scenario #1: Agro-Industry Oriented

289hm²	102hm²	106hm²
农业产业 Agro-Industry	休闲旅游 Leisure/Tourism	本地建设 Local Settlement

投资总额 Development Cost	¥3,013,000,000	创造就业 Employment	1279
流水 Turnover	¥1,963,440,000	游客数量 Visitor	4,080,000
人均收入 Per Capita Income	¥21,174	新增居民 New Resident	6360

场景计算 Scenario Calculation **面积 /** Area(hm²)	#1	#2	#3
农业产业 / / Agro-Industry	289	153	125
休闲旅游 / Leisure Tourism	102	195	116
本地建设 / Local Settlement	106	149	256
总计 / Total Area	497	497	497
开发成本 / Development cost			
农业产业 / Agro-Industry	¥2,601,000,000	¥1,377,000,000	¥1,125,000,000
休闲旅游 / Leisure Tourism	¥306,000,000	¥585,000,000	¥348,000,000
本地建设 / Local Settlement	¥106,000,000	¥149,000,000	¥256,000,000
总计 / Total Development Cost	¥3,013,000,000	¥2,111,000,000	¥1,729,000,000
年收入 / Annual Revenue			
农业产业 / Agro-Industry	¥1,734,000,000	¥918,000,000	¥750,000,000
休闲旅游 / Leisure Tourism	¥204,000,000	¥390,000,000	¥232,000,000
本地建设 / Local Settlement	¥25,440,000	¥35,760,000	¥61,440,000
总计 / Total Annual Revenue	¥1,963,440,000	¥1,343,760,000	¥1,043,440,000
就业 / Employment			
农业产业 / Agro-Industry	1156	612	500
休闲旅游 / Leisure Tourism	102	195	116
本地建设 / Local Settlement	21	30	51
总计 / Total Employment	1279	837	667
工资总额 / Wage			
农业产业 / Agro-Industry	¥78,030,000	¥41,310,000	¥33,750,000
休闲旅游 / Leisure Tourism	¥9,180,000	¥17,550,000	¥10,440,000
本地建设 / Local Settlement	¥530,000	¥745,000	¥1,280,000
总计 / Total New Wage	¥87,740,000	¥59,605,000	¥45,470,000
当前收入总额 / Total Existing income	¥21,746,400	¥14,225,600	¥11,342,400
增加的收入总额 / Increased Income	¥65,993,600	¥45,379,400	¥34,127,600
人均收入 / Per Capita Income	¥21,174	¥19,870	¥19,159
游客总量 / Tourist	4,080,000	7,800,000	4,640,000
居民 / Resident	6360	8940	15360

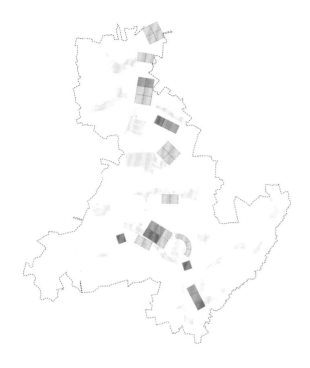

#2 旅游优先场景 Scenario #2: Leisure/Tourism Oriented

153hm²	195hm²	149hm²
农业产业 Agro-Industry	休闲旅游 Leisure/Tourism	本地建设 Local Settlement

投资总额 Development Cost	¥2,111,000,000	创造就业 Employment	837
流水 Turnover	¥1,343,760,000	游客数量 Visitor	7,800,000
人均收入 Per Capita Income	¥19,870	新增居民 New Resident	8940

#3 村屯优先场景 Scenario #1: Local Settlement Oriented

125hm²	116hm²	256hm²
农业产业 Agro-Industry	休闲旅游 Leisure/Tourism	本地建设 Local Settlement

投资总额 Development Cost	¥1,729,000,000	创造就业 Employment	667
流水 Turnover	¥1,043,440,000	游客数量 Visitor	4,640,000
人均收入 Per Capita Income	¥19,159	新增居民 New Resident	15360

一套更为灵活、更能迎接未来的规划系统

A PLANNING SYSTEM MORE FLEXIBLE AND READIER FOR THINGS TO HAPPEN

每个功能节点上都有三个功能类型可供选择：农业产业、休闲旅游、本地建设。根据不同的功能偏好或发展重点，可以有不同的组合，形成不同的功能用地分配。利用预测模型，不同的用地分配指标会产生不同的城市发展结果，包括投资策略、新增人口等。图示是三种根据不同功能偏好生成的应用场景及表现预测结果。

Each funcitonal node can choose from three functional types: agro-industry, leisure/tourism, local settlement. Different land use compositions can be formed based on different orientations and priorities in use. With the prediction model, each individual composition will yield different outcomes, including investment strategies, new residents, etc. The following is the application scenarios based on different priorities in use and their outcomes.

UCAP: UNROLLED CREATIVE

AGRICULTURAL PARK

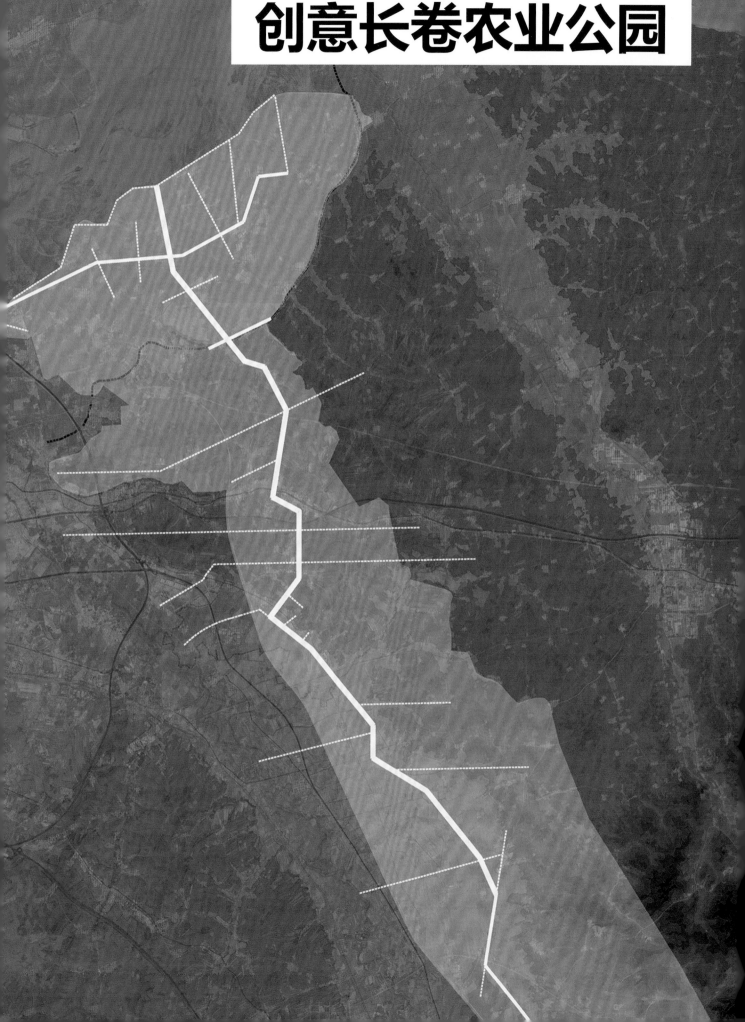

创意长卷农业公园

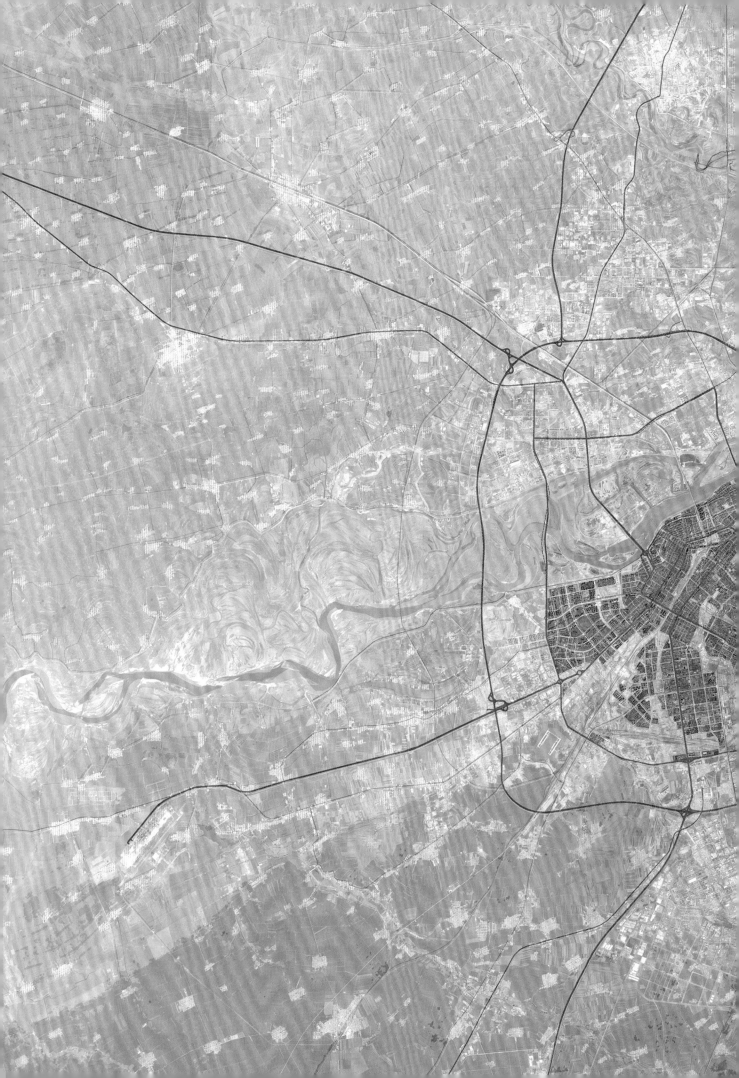

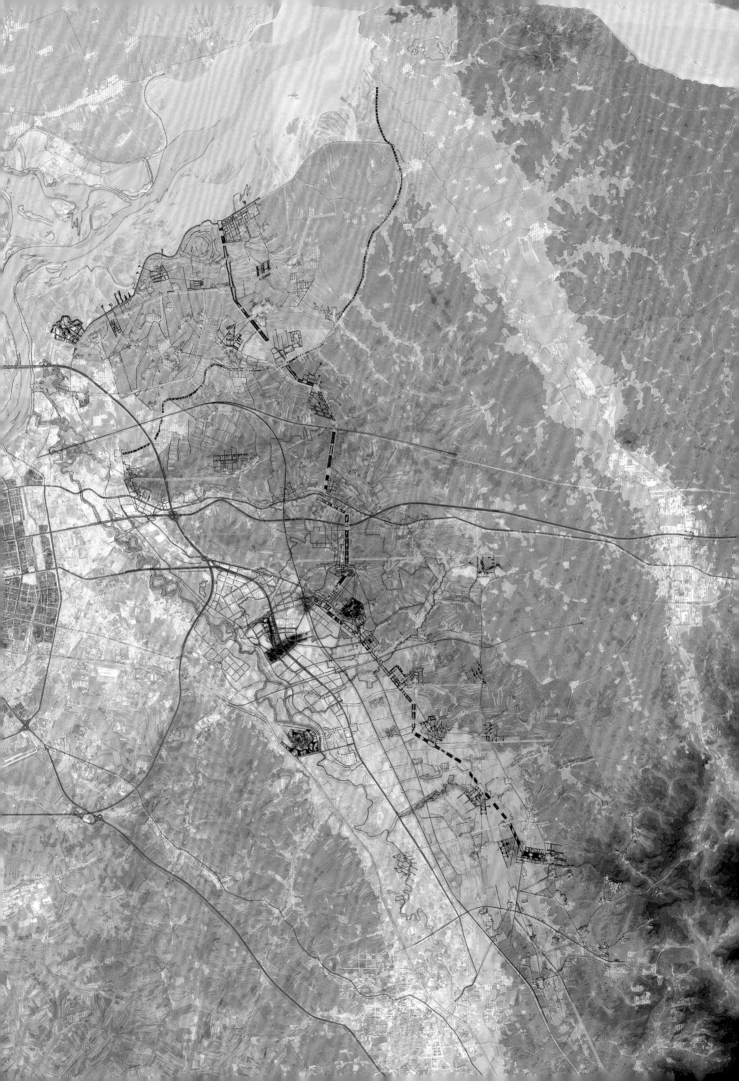

创意长卷农业公园总图规划：UCAP。

恢宏的提案远远超出了向阳的范围，并试图为这一雄心壮志的举措发展合适的规模：从松花江到山地；在阿什河走廊和另一个丰富多彩的蜚克图走廊之间。它可以延伸50km，平均宽度为0.8km。它可以从其中心即向阳项目"展开"。

基地非常靠近哈尔滨市郊，位于第四环路和阿什河绿色走廊的另一侧。由于这一位置优势，它具有集成的能力：

1. 新的商业活动。与农业产业相关，改革当前的生产实践和研究在极端气候条件下更好的农业生产。

2. 吸引游客。主要是通过为哈尔滨现有的声誉添加良好的品牌效应来满足国际需求，以及被各种基础设施主要是高铁的良好可达性吸引而来的大都市区和区域性的游客。

3. 通过新设施改善现有村庄并为住宅扩建创造空间。这是为了吸引渴望生活在创新农业环境中的大都市区的居民，并解决当前老龄化的问题。

这种设计方式为创意长卷公园提供了无与伦比的潜力，因为它具有混合不同经济类型和投资潜力的能力。由此看来，总体规划必须遵循特殊的逻辑；它需要比传统的总体规划蓝图更倾向"结构性的规划"：我们应该能够产生一个明确的"结构"——即极少的明确的形态规则——用于发展和改善整体景观质量的同时，确保已经在前文解释过的功能灵活性。因为它在本文件中被定义，这些将成为总体规划的主要优势。还补充了一些如何通过特殊的"机构"实施这一"结构性计划"，使主动行动成为可能的建议。

创意长卷农业公园的场地正在利用两个重要的价值或"优势"：土地的可用性、大都市区周边地理位置的非凡景观。

Masterplan for the "Unrolled Creative Agricultural Park" : the UCAP.

The proposal at the large scale is far beyond the boundaries of Xiangyang and tries to develop the right dimension for an initiative of that ambition: From the Songhua River to the Mountain; in between Ashi corridor and next affluent Feiketu corridor. It may extend 50km with an average width of 0.8 km. It can be "unrolled" from the center of it, that is Xiangyang Project.

Then the site is very close to Harbin metropolis and on the other side of 4th ring-road and Ashi green corridor. Because of its location it has the capacity of integrating:

1. New business activities: related to agro-industry, renovating current production practices and researching for a better agricultural production in extreme climate condition.

2. Atracting visitors, mainly from international demand by adding a good branding to existing Harbin reputation. But also, Metropolitan and regional visitors because of its good accessibility by different infrastructures and mainly by HST.

3. Improving current villages by new amenities and creating room for residential expansion, meaning attracting metropolitan residents eager to live in an innovative agricultural environment and solving the current aging problem.

This programing variety gives an unparallel potential for the UCAP because of its capacity of mixing different economical sectors and investment potentials. In this sense the Masterplan has to follow special logics; it needs to be more "structural plan" than traditional blue-print masterplan: We should be able to produce a clear "structure" – meaning few clear morphological rules– for the development and for the improvement of the quality of the general landscape and at the same time, ensuring the functional flexibility that is already explained in the "Matrix". Those are going to be the main strengths of the Masterplan as it is defined in this document. Complementary will be some suggestions how this "structural plan" can be implemented by special "Agency" that makes the initiative possible.

The site of the UCAP park is taking advantage of two mayor values or "strengths": availability of the land and extraordinary landscape in a metropolitan milieu and location.

公众轴线现今将江与山连到一起。
持续性的管理和组织能够在公园全境引入其他的活动

CIVIC AXIS CONNECTS MOUNTAIN AND RIVER WHICH, NOWADAYS, HAVE SOME INITIATIVES ON-GOING THAT WILL ADD OTHER ACTIVITIES TO THE OVERALL PARK

这些价值可以通过几个精确的"系统性元素"在模拟的"外围"中定义：

A　重组农业"田地"。这是发展的主要动力。在创意长卷农业公园，需要考虑新的种植方式。考虑到极端气候条件，这是一个需要深入研究现有开发可能的修改和转变的过程。然而，"田地"将是它的主要景观。此外，创意长卷农业公园的提升模式可以复制到哈尔滨东部和西部农村的网络中。

B　扩大城市架构。必须确保对创意长卷农业公园中包含的所有规划活动提供功能支持；高铁，可能的地铁线扩展和环路作为主要支撑。其余的将在向阳地区作为范例细化。

C　市民轴线和联合所有重要创新"节点"的理念。沿创意长卷农业公园构建中央绿色主轴。它的宽度约为100m，可能包含游客服务和公园活动支持。它的布局非常灵活，但它的方向是从山上"引导"到主河松花江。有些片段非常接近拟设的市民轴线，但不应被视为传统基础设施；相反，它是"绿色基础设施"，顺应地形和穿越基地并流往阿什河的水道；市民轴线是游客和居民共享的地方，可以包括河流和一些冰雕，并且在某些节点部分可以作为创意长卷农业公园活动的示范代表。

少量的垂直小分支可以连接主要市民轴线以外的其他节点。当接近松花江时轴线变成"T"形以支持与河岸更相关的正在建设中的节点；例如，为创意长卷农业公园的这一部分选择的活动将多是水上度假或健康养生、保健和治疗等。

D　选择不同功能的"节点"。正如之前章节所述，为给创意长卷农业公园和该地区创造经济活力，已经展开了对可能的经济活动的研究。它可以被讨论和丰富，但目前向阳区的转型过程——例如薰衣草公园和"温泉"度假区的开发——证明这个假设符合实际情况，但需要避免不受控制地占领农业用地，以及现有的和新建的基础设施之间的非协调使用。

当前和新的部分应包括用于主要城市空间和轴线的其他小型和不同用途的节点，避免封闭区域，寻找聚集公共空间并引入混合用途作为规划的一部分。

E　复兴现存村庄。它们是该地区的潜力，它们的品质提升可以改善现有居民的生活条件，并扩大公园内的住宅容量。

F　可持续环境的景观战略。合理地适应地形，通过平衡来保留大部分农业覆盖区未进行建造——不低于70%，可以确保产生积极的结果。

These values can be defined in the simulated "perimeter" by few precise "systematic elements":

A. Restructure the agricultural "fields". This is the main driver for the development. In the UCAP park, new forms of cultivation need to be considered. This is a process to develop with deep research on the possible modifications and transitions of the existing exploitations, taking into account the extreme climate conditions. Nevertheless the "fields" will be the main landscape for it. Also, improvements on the UCAP can be replicated into the network of agricultural villages in the east and west of Harbin.

B. Enlarging the urban armature. It must be ensured the functional support for all planned activities included in the UCAP; HTS, possible Metro extension, and Ring Roads as main supports. The rest will be detailed in the Xiangyang District as a sample.

C. Civic axis and concept for federating all important innovative "nodes". A central green spine is conceived along the UCAP. It has a width of approximately 100 meters and may contain services for visitors and supporting the park activities. Its layout is quite flexible, but it "navigates" from the mountain to the main river. Some sectors are very close to the proposed civic axis, but should not be seen as traditional infrastructure; on the contrary it is "green infrastructure" that acknowledges topography and lines of water crossing the territory and going to Ashi River; civic axis is place for visitors and residents, and can contain water and ice sculptures and in some parts of it can be representative of activities of the UCAP parks.

Few perpendicular small branches can connect other nodes that are aside of main civic axis. When approaching Songhua River axis turns into a "T" shape to support nodes more related to the riverbank, under construction; for instance, the activities selected for this part of the UCAP will be more water-resort or health, wellness and therapy, among other.

D. Selection of "nodes" of different programs. As described in former chapters, to create a dynamism into the economy for the UCAP and the region a research on the possible economic activities had been explored. It can be discussed and enlarged but the current process of transformation in Xiangyang District today-with Lavender Park and "Hot Springs" development, for example-prove that this assumption responds to a real condition, but it needs to be rationalized to avoid uncontrolled occupation of the agricultural land and a non-harmonic use of existing and new infrastructures.

Current and new sectors should include other small and different uses facing main civic spaces and axis, avoiding closed precincts, looking for gathering public spaces and introducing the mixed-use as part of the program.

E. Reactivation of existing villages. They are a potential for the territory and qualifying them may improve living conditions of current residents and extend the residential capacities within the park.

F. Landscape strategy for sustainable environment. The right adaptation to the topography and good balance to keep large proportion of the agricultural coverage unbuilt-meaning no less than 70%-can ensure this positive outcome.

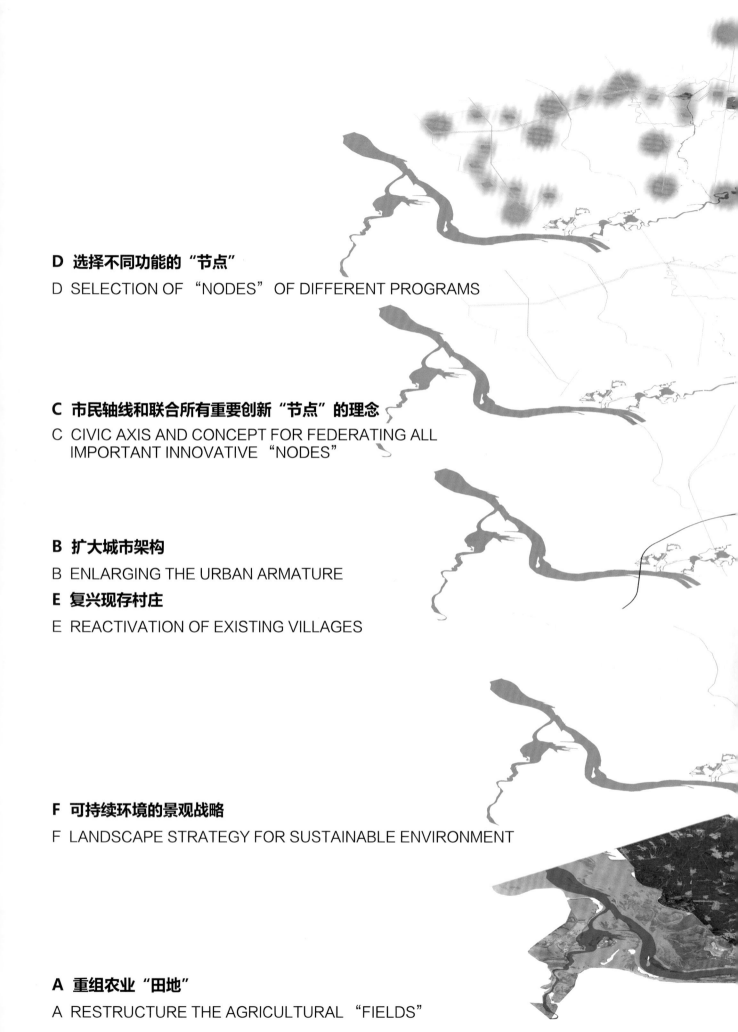

D 选择不同功能的"节点"

D SELECTION OF "NODES" OF DIFFERENT PROGRAMS

C 市民轴线和联合所有重要创新"节点"的理念

C CIVIC AXIS AND CONCEPT FOR FEDERATING ALL
IMPORTANT INNOVATIVE "NODES"

B 扩大城市架构

B ENLARGING THE URBAN ARMATURE

E 复兴现存村庄

E REACTIVATION OF EXISTING VILLAGES

F 可持续环境的景观战略

F LANDSCAPE STRATEGY FOR SUSTAINABLE ENVIRONMENT

A 重组农业"田地"

A RESTRUCTURE THE AGRICULTURAL "FIELDS"

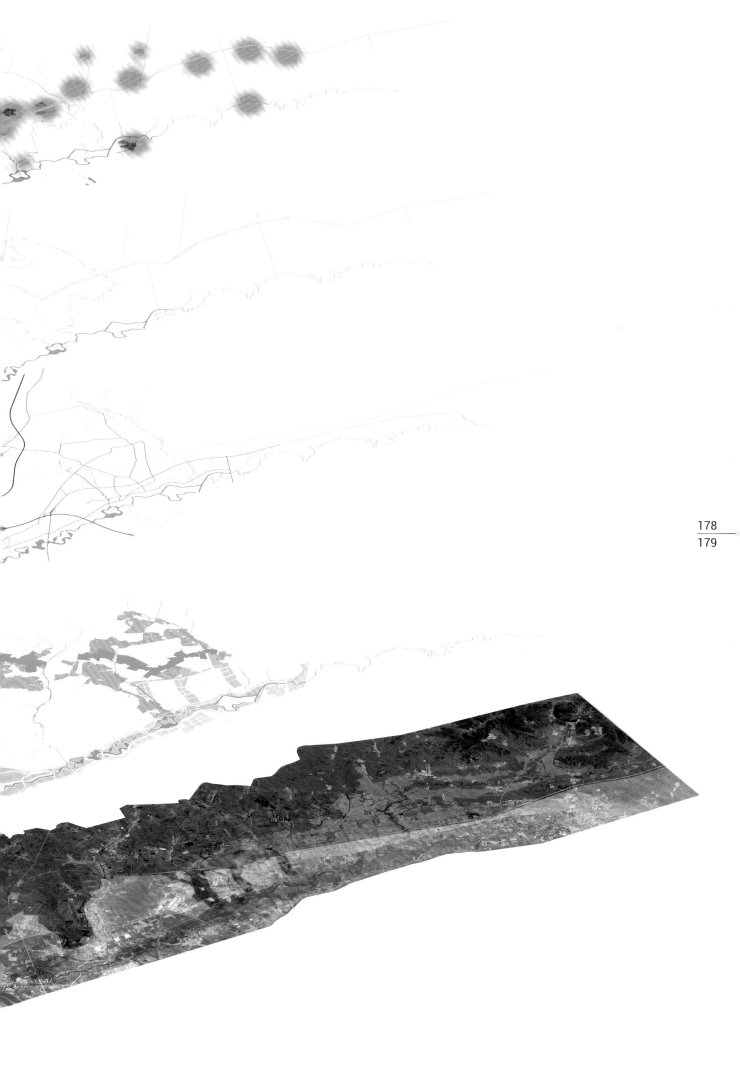

哈尔滨团队到访巴塞罗那（2018年6月）
Harbin's team visits Barcelona (June 2018)

布斯盖兹教授在哈尔滨市政府的讲座（2018年10月）
Prof. Busquets' lecture at Harbin People's
Government (October 2018)

[1] BOOMEN T V, et al. Urban challenges, resilient solutions: Design thinking for the future of urban regions. Haarlem: TrancityxValiz, 2017.

[2] GALAND G, AUTISSIER I, GERMA P. La ville renaturée: Réconcilier lespace urbain et la biodiversité. Paris: Éditions de La Martinière, 2015.

[3] GIROT C. The course of landscape architecture: A history of our designs on the natural world, from prehistory to the present. London: Thames & Hudson, 2016.

[4] Harbin yin · xiang. Beijing: China Architecture & Building Press, 2005.

[5] HOLMGREN D, Kaïm A E, COCHET Y. Permaculture: Principes et pistes daction pour un mode de vie soutenable. Paris: Rue de léchiquier, 2017.

[6] LAMBERT E, Lévy J. Le parc planétaire: La fabrication de lenvironnement suburbain. Paris: LOeil dor, 2018.

[7] MATHIS C, Pépy E. La ville végétale une histoire de la nature en milieu urbain (France XVIIe–XXIe siècle). Ceyzérieu: Champ Vallon, 2017.

[8] RANZATO M. Water vs. Urban scape: Exploring integrated water-urban arrangements. Berlin: Jovis, 2017.

[9] SCAGLIONE P, Breda B. Cities in nature: Ecourban design, landscapes, slow cities = Città nella natura. Trento: LISt, 2012.

[10] SERY J, SAUNIER F. Ruralités et métropolisation: à la recherche dune équité territoriale. Saint-Etienne: Publications de lUniversité de Saint-Etienne, 2016.

[11] SOMMARIVA E. Cr(eat)ing City: Strategie per la città resiliente. Trento: LISt Lab, 2014.

[12] STEINER F R. Making plans: How to engage with landscape, design, and the urban environment. Austin: University of Texas Press, 2018.

[13] Urban Planning Bureau of harbin Municipality and Urban Planning Society of Harbin Municipality. Glance back the old city's charm of Harbin. Beijing: China Architecture and Building Press, 2005.

[14] Harbin Urban Planning Bureau. Central Avenue: Memorial stamp collection of the awarded regeneration project for historic block of central avenue, 2005.

[15] ZEEUW H D, DRECHSEL P. Cities and agriculture: Developing resilient urban food systems. Basingstoke: Taylor & Francis, 2015.

布斯盖兹教授的著作
PROF. BUSQUETS' BOOKS

多元路线化都市.
武汉:华中科技大
学出版社，2010.

*Cities: X Lines.
A new lens for
the urbanistic
project*. Harvard
University
Graduate School
of Design, 2006.

杭州：从运河
网络到超大街
区，ORO版.
2017.

*Hangzhou
underlays. Grids
from canal to
maxi-block*. Oro
Editions. 2017.

深圳：设计不停变化
的城市. 哈佛大学
设计学院，2010.

*Shenzhen:
Designing
the non-stop
transformation
city*. Harvard
University
Graduate School
of Design, 2010.

巴塞罗那：全尺度的
都市规划之路. 建
筑师，
2018，191.

*Barcelona: All
scale road
for urban
planning*. The
Architecture,
2018, No191.

城市建筑设计：胡安•
布斯盖兹+BAU/
BLAU. 世界建
筑，2013，272.

*Designing urban
Architecture:
Joan Busquets+
BAU/BLAU*. WA
Magazine, 2013,
No272.

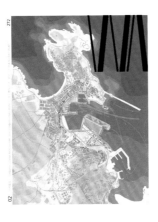

重庆：寻求秩序性
作为当代城市设计
的革新模式. 哈佛
设计学院，2018.

*Chongqing:
Searching for
regularity as a
transformative
model in the
design of the
contemporary
city*. Harvard
University
Graduate School
of Design, 2018.

巴塞罗那：一座紧凑
城市的城市演变.
北京：中国建筑工
业出版社，2016.

*Barcelona: The
urban evolution
of a compact
city*. San
Raphael: ORO
editions, 2014.

图书在版编目（CIP）数据

绿道：哈尔滨向阳镇发展战略规划 = Greenway——
Development Strategic Planning of Xiangyang Town
in Harbin：英汉对照 /（西）胡安·布斯盖兹
（Joan Busquets）等著 . —北京：中国建筑工业出版社，
2020.3

　　ISBN 978-7-112-25763-8

　　Ⅰ.①绿…　Ⅱ.①胡…　Ⅲ.①城镇—城市规划—哈尔
滨—英、汉　Ⅳ.① TU984.235.1

　　中国版本图书馆 CIP 数据核字（2020）第 256223 号

责任编辑：焦　扬　陆新之
书籍设计：康　羽
责任校对：王　烨

绿道——哈尔滨向阳镇发展战略规划
GREENWAY：Development Strategic Planning of Xiangyang Town in Harbin
[西] 胡安·布斯盖兹（Joan Busquets ）　高　岩　周艳莉　周雪瑶　著
*
中国建筑工业出版社出版、发行（北京海淀三里河路 9 号）
各地新华书店、建筑书店经销
北京雅盈中佳图文设计公司制版
天津图文方嘉印刷有限公司印刷
*
开本：880 毫米 ×1230 毫米　1/16　印张：11¹/₂　插页：1　字数：414 千字
2022 年 1 月第一版　2022 年 1 月第一次印刷
定价：88.00 元
ISBN 978-7-112-25763-8
　　（37000）